JN418470

From Biotechnology To Bioindustry

From Biotechnology To Bioindustry

Seung Wook Kim

Translated by Kyung Yeon Kim & Seung Wook Kim

KOREA
UNIVERSITY
PRESS

Korea University Press
145, Anam-ro, Seongbuk-gu, Seoul, 02841 Republic of Korea

ISBN: 978-89-7641-989-7 93500

Printed in Republic of Korea

This is a revised translation of the original Korean Edition.
생명공학기술과 바이오산업. 2016 by Korea University Press, Seoul, Republic of Korea.

No absolute fact exists within the ones written on literatures and texts. All facts are still ongoing topics under research. Curiosity and asking questions bring another inception. A sustainable trial gives you creativity. To create something, recreation is important. Enjoy your research and recreation.

-Seung Wook Kim

PREFACE / AUTHOR'S NOTE

Health, energy, and environment fields have been growing immensely in the recent years within the twenty-first century. The fast growth in these fields was achieved because of increased interest and investment by many different countries due to its close relation and high demand in national security. With the growth of these fields, some of the related fields have also been growing exponentially, such as biotechnology, information and communication technology, nanotechnology, energy and environment technology, space technology, and cultural technology. In the current state, each of these is quickly converging together to build new and innovative technologies. Out of all, biotechnology has especially been playing an important role as a bridge in creating modern technologies through convergence with other fields.

Bioindustry and bioeconomy have also been consistently and actively developing with the foundation of biotechnology. Therefore, I thought understanding the basis of biotechnology and multifarious concepts of bioindustry was important for everyone who are living in the modern days. As a result, I have decided to organize and write this book on the basis of biotechnology and bioindustry that I have been teaching for many years at Korea University. As the title of this book 'From Biotechnology To Bioindustry' shows, I tried to draw out the concept of how biotechnology is applied in the field of bioindustry as simple as possible.

Chapter 1 briefly explains the development of technology and industry field, and Chapter 2 talks about the basic concept of bioresources, which are the most important component of bioindustry. Then, Chapter 3 and Chapter 4 contain the history, functions, and application of the key elements of bioresources – cells and enzymes. Chapter 5 describes the basic bioprocesses and production of various biotechnological products with the use of cells and enzymes. Chapter 6 talks about the types and characteristics of bioreactors that yield biotechnological products, and Chapter 7 talks about immobilization technology for efficient bioconversion. Then, Chapter 8 explains the method of separation and purification of the final product in biotechnology. Chapter 9 mainly focuses on the application and overall trend of medical biotechnology field and products that use bioresources. Lastly, Chapter 10 explains the basis and application of industrial biotechnology based on the biomass which is the future bioresource to be used.

I would like to express my gratitude to Mr. Soo Kweon Lee, Mr. Ji Hyun Yang, and other graduate school students of Bioprocess Engineering Laboratory, the Department of Chemical and Biological Engineering at Korea University for helping with pictures and graphs. Also, I would like to thank Dr. Hunsa Punnapayak, Dr. Sehanat Prasongsuk, and Dr. Pongtharin Lotrakul from the Department of Botany at Chulalongkorn University for letting me finish this book during my sabbatical year in Thailand. I appreciate consistent love and courage from my wife Eun Kyung, my daughter Kyung Yeon, and my son Yusik, and I want to dedicate this book to my father who is now in heaven. Also I would like to give many thanks to my daughter, Kyung Yeon, who made a dedication for the publication of an English version of this book.

At last, I am sincerely thankful to the Korea University Press and Institute of Basic Education of Korea University for helping me to publish this book.

2018. 12 from Anam,, Seoul, South Korea
Seung Wook Kim

CONTENTS

Chapter 1
Industry, Technology, and Bioindustry

In the modern times, knowing the basic correlation between science, engineering, technology, and industry is vital to understand the general economic development of a country. Industry that builds the foundation on economy does not occur naturally, but rather originates from the basic science knowledge. Science mentioned in here includes not only the natural science but also the studies such as cultural and social sciences. In general, governments or companies invest on education and research on science and engineering fields depending on the demand and necessity at that moment. Through these investments, many achievements can be obtained. For science, obtaining new and fundamental knowledge is crucial whereas engineering must focus on finding and developing new technologies with the application and utilization from the facts that were discovered from the basic sciences. Engineering field can emerge and introduce new diverse industries by consistently developing new technologies. During this process, the importance to focus is the technology transfer with the basis of IPRs (Intellectual Property Rights). As for IPRs, domestic and international patent applications must be prioritized before anything else, and is usually valid for 20 years in South Korea, the United States, and the EU nations. A very good example of the patent war that happened recently is the confrontation between Samsung and Apple with their smartphone-related technologies. As of right now, many companies are reinforcing professionals and experts to strengthen and prepare the IPRs dispute and any future patent wars. In particular, preparations against the professional patent management companies also known as 'patent trolls', have also been very intimidating. Through the technology transfer with

Figure 1-1 Basic relationship between the industry and science/engineering/technology

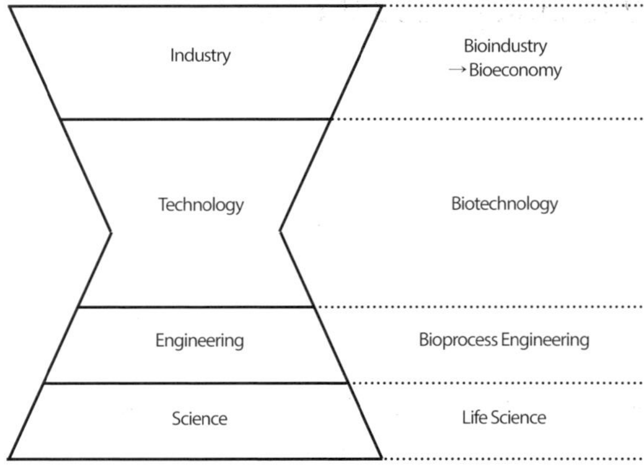

the basis of patent, industrial activities of the companies generally become more dynamic, which subsequently improves the economy of a country. The prime and fundamental importance during these series of processes is to discreetly consider the aspect of humanities that every work is for the human well-being (Figure 1-1).

The history of an industrial development is very crucial to any countries. By looking back into the past, we gain the wisdom to understand the present duties as well as minimalizing any failures. Although it is hard to predict the future from history, studying the history can give motivation and hope to grow for future creations. For instance, in 1960s, Korea was in serious poverty with the gross national income per capita (GNI) of USD 79. However, continuous plan development for economic policies and systematic development by the Korean government from the 1960s to 1980s have saved Korea to get out of poverty. Starting from 1962 with the First Economic Development Five-year Plan, South Korea promoted economic development with the basis of textile industry. Especially in the 1970s, intensive cultivation on heavy chemical industries such as steel, petrochemical, automobile, machinery, shipbuilding, and electronics has built the foundation for today's economic development.

Though many economic crises had occurred such as the first and second oil shocks in the 1970s and 1980s, and the labor-management dispute

Figure 1-2 History of industrial growth in South Korea and increment of GNI per capita ($(year))

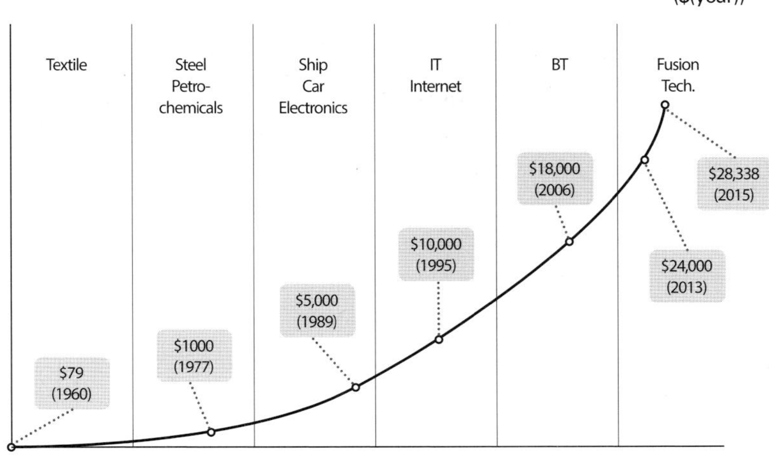

and the foreign exchange crisis in the late 1980s and 1990s, the South Korean economy survived throughout these hardships with continuous cultivation of heavy chemical industries as well as the emergence of major corporations. Plus, rapid industrial structure changes had happened as some of the major corporations started launching into the semiconductor industry. As semiconductor industries and other various manufacturing businesses consistently grew, gross national income per capita has also increased. By 1995, GNI per capita reached to USD 10,000. Twelve years later, in 2007, GNI per capita reached to USD 20,000 for the first time, then USD 28,000 by the year of 2015 (Figure 1-2).

As a reference, national income represents the improvement of a country's level of living and economic activity in comparison to the previous year, and is generally shown in combination with the added value during the whole one year. Normally, two types of indicators are used to represent the national income quantitatively – Gross National Income (GNI) and Gross Domestic Product (GDP). GNI per capita is defined as the sum value of income claimed by all residents in a nation. In here, all income received from both domestic and foreign nations are included, but does not include the output paid to the non-residents of the country. GDP is defined as the sum of total value of national goods and services produced by a nation, and includes all domestic production.

As of right now, various important and fundamental elements that have affected on the nation's growth in industries must be carefully considered and valued. In addition, continuous observation and analysis on possible factors that may play an important role in the future competitive global societies must be made. In regards to this perspective, South Korea did a great job with building different characteristics in their industry. Starting from 1980s, South Korea had adapted well with the change of paradigm in centering onto the product and technology development instead of being dependent on the systems and policies given by the government. Starting from 2010, South Korea has been showing numerous potentials for the future by entering the paradigm of an advanced country with the focus on the society and culture. A good illustrative case example is the iPhone where Korean industries were in jeopardy by falling behind the global trend because of their strong mindset oriented on manufacturing new products. Even though the Korean industries had gone through such a jeopardy, they could overcome the situation and are now in strong competition against other developed countries. This result was expected due to the continuous education to produce more professional experts and tight cooperation between industry, academy, and administration, which will be explained later in detail.

Therefore, it is necessary for various South Korean industries to stand with the mindset of a developed country to have businesses shoulder-to-shoulder and compete against the industries of developed countries. Through this paradigm change, the idea of the 'Creative Economy' can be applied with increased growth engine. Although many specialists and experts are concerned that something may be wrong with the growth engine in Korea, growth engine is not something that can be newly built from nothing. Rather, it is a process of constructing higher value-added industries by discovering innovative industries with increased professional workforce based on the traditional manufacturing practices that have been existing and building for many years.

A good example of this would be an excellence in Korean speed skating skills. As recent Korean speed skating athletes are significantly excelling in

Figure 1-3 Relationship and integration of science, engineering, technology and industry

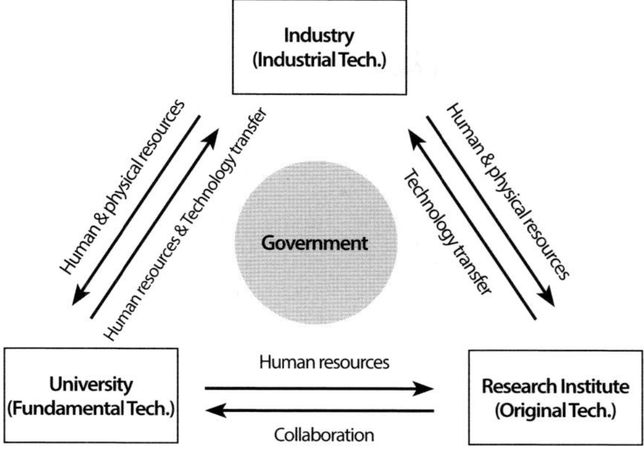

the Winter Olympics, this progress is believed to be the result of continuous investment and active participation in sports sciences. More specifically, the experts believe that the success in Korean speed skating came from the long-term and sustainable investment, improvement in basic physical fitness, advanced uniform materials with reduced air resistance, and putting short track speed skating skills all together in application. As shown, speed skating has shown such success in the Winter Olympics due to the integration of the science, engineering, technology, human resources, and investment.

Similar to the example above, development on industries must establish a successful system that enables the operation between science, engineering, technology, and the industry integrated together. To build such an integrated relationship between these various fields, the government, businesses, universities, and research institutes must work hard not only on individual growth but to have a strong partnership with each other. Moreover, it is crucial to cultivate the groundwork for the medium and large business companies to unite together to give more support to small and medium businesses instead of letting only the large business to grow larger (Figure 1-3).

Regardless of the different fields, history shows that diversity is the most important factor in the growth of an organization or a nation. In 1840, the Great Famine occurred in Ireland was caused by the outbreak of potato blight across the whole country. From the outbreak of potato blight, around one million people lost their lives, and several million people migrated to America. One of the causalities of this high mortality rate and migration rate was due to decreased number of potatoes, which was their staple food. Although part of the reason to the decreased number in potatoes was because of exploitation by the Great Britain when Ireland was colonized under them, the main reason was due to the selective and monoculture breeding of potatoes for the highest quality and productivity, which these potatoes lacked resistance against potato blight. As described in this story, diversity and preservation of biological resources are very important. An interesting fact is that Henry Ford, the founder of the Ford Motor Company, Richard and Maurice McDonald, the founder of McDonald's, and John F. Kennedy are all decendents of the Irish immigrants. Without the potato blight outbreak, these famous people would not have existed in the United States. The Irish Potato Famine directs us to an importance in the genetic diversity of potatoes.

Another example is Nokia, a renowned company from Finland. Nokia started as a paper manufacturing company in 1865, and grew as one of the most prominent IT companies in the world as they started focusing more onto the IT field in the 1990s. Despite the significant growth, Nokia ended up being taken over by Microsoft. The failure of Nokia was due to losing reputation with lack of investment in research and development, and inability to adjust to the quickly changing global IT field as Samsung and Apple companies started emerging with newly-developed technologies. However, a fortunate part for the country of Finland is that about 300 new startup businesses were established by the workforce from Nokia by the time they were diminishing. Nokia is not the only company that had gone through this problem. Many large IT companies such as Sony, Hewlett-Packard, or Kodak, are also suffering or shrinking or closing down with similar reasons. Through the observation, we can learn that medium and large-sized enterprises must cooperate and read through the future patterns and changes, and continuously make an effort to develop and grow in order to survive in such a competitive global economic period.

In the current times, six technologies are consistently and extensively

Figure 1-4 Fundamentals of fusion technology

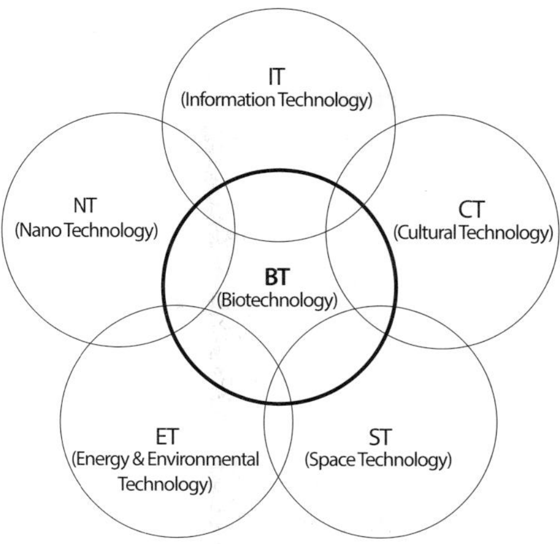

developing under extensive research throughout the whole nations: biotechnology (BT), information technology (IT), nanotechnology (NT), energy and environment technology (ET), space technology (ST), and culture technology (CT). Each of these technologies is rapidly progressing individually, and merging together to create new innovative technologies. Integration of these different technologies is called Fusion Technology or Convergence Technology. An example of fusion technology is BINT, which is a convergence of biotechnology, information technology, and nanotechnology. A protein chip, which is a part of Nano-biochip, is made through BINT. Because the protein chips are used in IT field, where biological materials of protein and enzymes are combined with the semiconductor board, it is considered as a typical fruit of convergence technology (Figure 1-4).

Actually, South Korean government started supporting biotechnology since 1982, as the Ministry of Science in South Korea selected biotechnology as the Core Technological Strategy. As Genetic Engineering Promotion Act was established in 1983, many genetic engineering majors and degrees were established throughout many universities. In present, most of the universities changed the name of genetic engineering into either biotechnology or biological sciences program. In addition, Genetic

Engineering Center (now called Korea Research Institute of Bioscience and Biotechnology) was established for further support for biotechnology.

Concept of biotechnology builds its foundation upon the manufacturing process for various biotechnological products with utilization of microorganisms, animal cells, plant cells, and enzymes by universities, research institutes, industries, and the cooperation between them in variety of fields. In general, successful biotechnological products can be made only when there is a harmony in between upstream and downstream technologies. Upstream technologies have their basis on basic sciences such as biology, microbiology, biochemistry, molecular biology, and molecular genetics. These types of technologies normally study on organisms and their biological mechanism and genetic information. On the other hand, downstream technologies are built upon biochemical engineering and are focused mainly on the bioprocess. Bioprocess can be divided into three different processes: strain development, mass production of the product, and separation of the product. As each of these bioprocess fields has been under active research, the field of bioindustry has grown immensely with the creation of bioeconomy.

Although biotechnology includes many different fields of technologies, it is mainly divided into four different areas based on different colors: Red Biotechnology, White Biotechnology, Green Biotechnology, and Blue Biotechnology. Red Biotechnology represents Medical Biotechnology as the color of red symbolizes blood, and is the field of health including biomedicine and bio-organs. White Biotechnology represents Industrial Biotechnology which produces biofuels, white-colored biochemical products, and includes bioprocess and bioenergy fields. Green Biotechnology is the agro-livestock field that controls over green plants including living modified organisms (LMOs), genetically modified organisms (GMOs), and health supplements. Lastly, Blue Biotechnology is imagined as blue ocean, therefore, is referred as Marine Biotechnology. The ocean consists of vast amounts of bioresources, which can be utilized with endless potentials (Figure 1-5).

With the development of biotechnology in the 1970s and the production of numerous bio-products in the 1980s, bioindustry started getting its attention by the time of 1990. Bioindustry, also called as biotechnol-

Figure 1-5. Four main biotechnology areas in colors

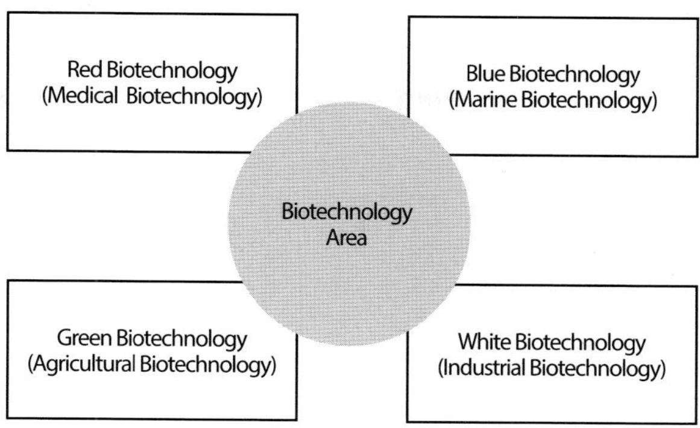

ogy industry, is an industry that produces various high value-added biotechnological products by using bioresources with the knowledge of biotechnology. Biotechnology can be divided into conventional and modern biotechnology depending on the presence or absence of genetic recombination technology. Conventional biotechnology does not mean it is a boring old-schooled technology. Addition of modern technology, such as gene recombination, from the conventional technology had developed and flourished all fields of biotechnology with emergence of many new technologies.

On February 12th, 2001, the scientists in the world announced that they have found over 99% of human genome that is made of over three billion base pairs. This important fact was found through Human Genome Project (HGP) that was conducted by over ten countries such as the United States, Japan, Germany, and the United Kingdom, with the cooperation of a private company called Celera Genomics. The scientists successfully completing the human genome map gave a massive potential for growth of bioindustry field with the understandings and application of genetic mechanisms. In here, the word ‘genome’ is a portmanteau of ‘gene’ and ‘chromosome’, and includes all genetic sequences and information of an organism.

With the big success that Human Genome Project has brought to the future of human beings, a new era of Post-Genome is coming for further research and application of genome. As of right now, the researchers are

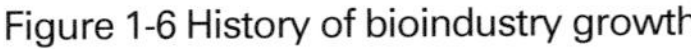

Figure 1-6 History of bioindustry growth

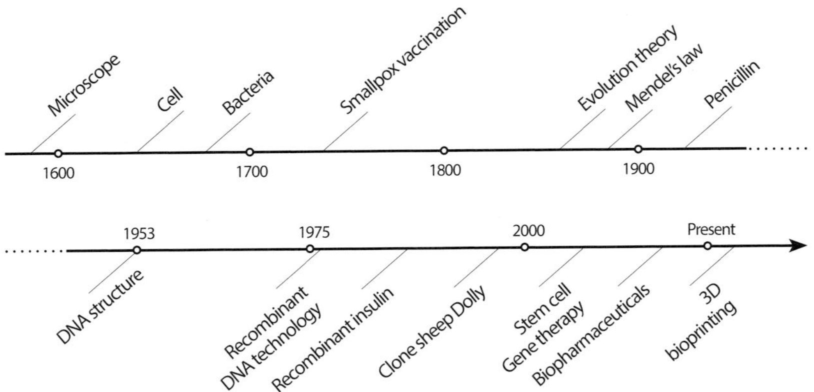

focusing on gene therapies, somatic-cell nuclear transfer therapies, and stem cell technologies to treat human diseases. Therefore, bioindustry is expected to bring a massive ripple effect towards energy, environmental, medical, pharmaceutical, agricultural, and electronical industries, which is a lot more than what the IT industry growth had brought within the last decade. As a reference, Figure 1-6 shows the history of bioindustry from the past to present.

Ultimately, the most important part for the industry to grow is to consistently support and raise many bright future manpower who are studying different science and engineering programs. With this future workforce, many new and creative technologies should be built in order to produce high value-added products in the future. Bioindustry field is especially an important area that needs major financial investments since it is an easier field to be integrated with other fields of industries. As mentioned, bioindustry and the use of bioresources in the field are crucial factors to be considered in order for the creation of advanced future technologies. Next chapter is going to talk about the basic concept of numerous bioresources.

Chapter 2

Basic Concept and Application ofVarious Bioresources

Bioresources can be applied in many different areas of bioindustry with their infinite possibility of use. Bioresources consist of all living organisms, all the parts that make up organisms such as DNA, and all the information and functional materials made by the organisms. These bioresources have immense possibility of application in many different fields. The word 'BIORESOURCES' contain many different meanings to it:

1) B: Biological Resources

Bioresources indicate organisms that are alive, their compositions, and the materials or information that are produced by these organisms.

2) I: Industrial Application

If the high value-added products were produced by using bioresources in an industrial scale, that can give quite a lot of benefits to the human beings.

3) O: Organisms

It is important to confirm and discover the existence of diverse organisms living in a nation and to gather much information about various organisms that reside in other nations.

4) R: Reuse

Bioresources that utilize living organisms can always be reused.

5) E: Energy

Many organisms continuously produce energy, and some organisms use themselves as an energy source.

6) S: Save

Continuous accumulation of information on bioresources and their functions become an important asset to a nation.

7) O: Originality

Any novel materials or functions can be found from almost all bioresources depending on the different purpose of use.

8) U: Usefulness

Considering the newly invented products or machines that mimic bioresource functions, one cannot emphasize too much on the importance of bioresources.

9) R: Reconsideration

Although many products and machines are developed with current bioresources, the potential of developing better and higher value-added products and machines is limitless. Therefore, it is always important to reconsider and emphasize on the importance of these factors.

10) C: Conservation

Conservation of bioresources is such a critical factor to consider. As the nature and the environment have gotten destroyed and polluted, we have already lost so many bioresources. This means the nation has lost such valuable treasures.

11) E: Evolution

Over time, all living organisms have gone through evolution and selective pressure at a slow rate according to the changing environment of the Earth. However, the current technologies enable humans to purposefully modify genes for the organisms to go under evolution at a fast rate for a specific purpose.

12) Sustainability

Sustainability is one of the most important fundamental concept of bioresources, and the genetic information and its industrial application are continuously applicable. The fact that bioresources have 'sustainability' gives an importance in energy, environmental, and public health care fields as eco-friendly resources.

As shown above, there are many different meanings behind the word 'bioresources'. As of right now, microorganisms, plants, animals and their genetic materials, enzymes or functional materials produced by these organisms are the most important bioresources. Recently, human-derived biological resources have been considered as another important bioresources.

As mentioned above, bioresources can be found in many different parts of organisms. Microorganisms include important bioresources such as bacteria, archea, fungi, viruses, microalgae, mushrooms, and cordyceps. In plants, different seeds, plants themselves, their tissues, the plant cell lines, and transgenic plants can be used as good bioresources. Animal

Figure 2-1 Various kinds of bioresources

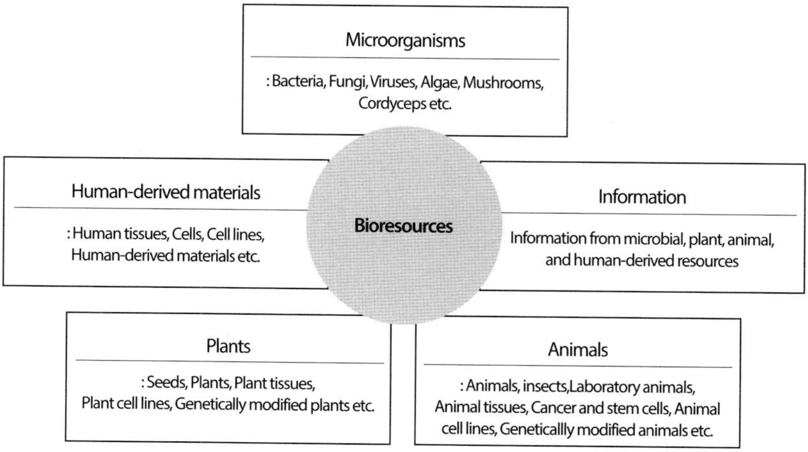

Microorganisms
: Bacteria, Fungi, Viruses, Algae, Mushrooms, Cordyceps etc.

Human-derived materials
: Human tissues, Cells, Cell lines, Human-derived materials etc.

Bioresources

Information
Information from microbial, plant, animal, and human-derived resources

Plants
: Seeds, Plants, Plant tissues, Plant cell lines, Genetically modified plants etc.

Animals
: Animals, insects,Laboratory animals, Animal tissues, Cancer and stem cells, Animal cell lines, Geneticallly modified animals etc.

bioresources include animals themselves, laboratory animals such as insects, mice, or primates, animal tissues, cancer cells and stem cells, and hybridoma such as animal cell lines, fertilized egg, transgenic animals. Also, human-derived bioresources such as human tissues, cells and cell linings, human-derived materials and information can be used as long as they abide by the law of bioethics (Figure 2-1).

If we are able to efficiently use these various bioresources, we can produce high value-added products that can increase a wealth of a nation. In 1999, an American microbiologist David Perlman (1920 – 1980) included some interesting facts about microorganisms in the journal titled as "Laws of Applied Microbiology", which was published by the International Journals called Journal of Industrial Microbiology & Biotechnology. Perlman is claiming that the future would be a lot more promising if the humans learn how to use the microorganisms efficiently. The truth is that less than 10% of microorganisms that exist on Earth are being utilized. Continuous and more proactive research must be operated for better untilization on not just land and oceanic microorganisms, but also on the animal and plant cells.

Agriculture, medicine, environment, energy, and food industries are some of the important applied fields that use the bioresources mentioned above. Each of these fields is currently actively applying these bioresources on use. Let's look at some examples:

Though large number of crops are being produced in the agricultural field, these crops are being applied with strong pesticides due to the problem of pests damaging the crops. The popular use of pesticides is speeding up the destruction of ecosystem with much pollution. Not only that, the farmers use herbicides against over-growth of weeds on the cultivated fields. The only problem is that the crops are quite prone to be killed by these herbicides. There are other problematic situations such as climate changes causing reduction or impossibility of crop production due to air pollution. In order to face these rising problems, the scientists have been trying to find new ways to make crops to be resilient against herbicides or to grow easily under lower temperatures through genetic recombination.

Genetic recombination technology has its strength in improving the diverse breeds of organisms at a short period of time by inserting a specific gene of interest into the host cell. This can lead to expansion in crop production and reduction of pesticide use. The crops that are cultivated by this technology are called genetically modified organisms (GMOs). The GM foods that are produced by genetically modified crops have had high disputes with safety at first, but had decreased after the stability analysis and approval of GMOs by FDA in 1994. Ever since then, many different GMO crops such as corn, beans, or wheat have been produced. As a reference, living modified organisms (LMOs) is a generic term that includes all GMO and microorganisms.

The traditional method of cross-breeding the similar breeds of crops with desired characteristics is still a popular area of research, but this traditional breeding system has a drawback of having high trial and error, and long process time. However, the breeding period and the expenses has shortened significantly after the year of 1980 when the researchers started using the selection technology with DNA markers. After the mid-1990s, breeding between the GMOs had gotten more popular as genetically modified crops started to become commercialized. Plus, new breeding systems have developed with the use of genome or metabolome as biotechnology has developed.

Another example to increase the crop yield is to use nemesis of the pests. This natural enemy decreases the number of pests by either attacking and killing the pests or by limiting its sexual reproduction. These types are called biopesticides, which use 1) non-human-infecting virus that is only virulent to pests; 2) nematodes; 3) microorganisms; 4) toxins that are produced by microorganisms towards pests. These biopesticides either give diseases to pests, or kill pests with specific toxins. Biopesticides do not cause environmental pollution, therefore, the research towards biopesticides are quite active and many are already commercialized to protect crops from variety of pests (Figure 2-2). As explained above, specific genes that are used in GMOs, natural enemies, viruses specific to pests, microorganisms and their toxins are very important bioresources. Especially, some types that only exist in certain nations should be highly and strictly protected by the nation. Nowadays, some use mixture of normal pesticide and biopesticide to decrease the use of normal pesticide. In this case, biopesticides must be inserted with a pesticide-resistant gene through genetic modification or find a way for it to keep its tolerance towards pesticides to prevent the organisms or materials that are used in biopesticides from losing their ability by the normal pesticides.

Figure 2-2 A grasshopper infected by fungus

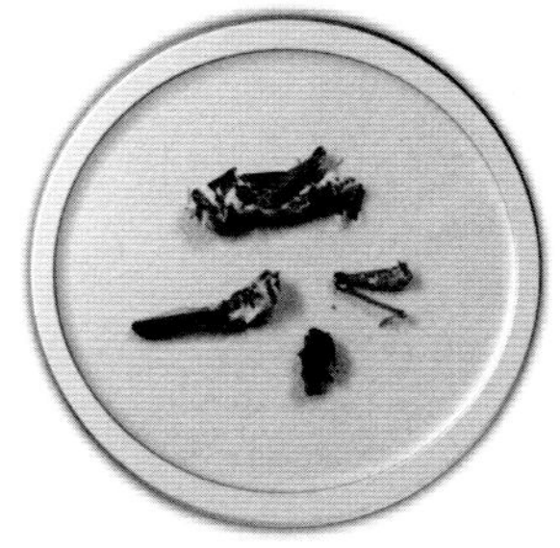

One of the examples in the medicine field is penicillin. Since penicillin was discovered, many different types of antibiotics have been found throughout the years. From these antibiotics, many other new antibiotics were synthesized via chemical and biological methods. In general, antibiotics can inhibit or completely kill the growth and spread of pathogens that are involved in human diseases with very small amount. This concept and idea became more organized as Alexander Fleming (1881 – 1955), an English bacteriologist, discovered penicillin in 1928. Fleming mainly researched on the characteristics of Staphylococcus. When he found the world's first antibiotics during his research, he decided to call it 'penicillin'. Fleming was going on a summer vacation for two weeks with his family, so he decided to put the petri dish that was cultured with Staphylococcus on the lab bench. When he came back from the vacation, he found that the Staphylococcus plate was contaminated with fungus that appeared to be yellow and green. Later, Fleming found that the fungus that contaminated the plate was a *Penicillium notatum* spore that flew in from a fungal laboratory downstairs. When Fleming examined the contaminated plate, he could not find any Staphylococci growth on the areas where *P. notatum* had developed. Also, Staphylococci that grew around *P. notatum* had a distinct line, which drew a perfect circle. With these factors he observed, Fleming hypothesized that *P. notatum* may produce materials that may be inhibiting the growth of bacteria. This scientific finding was possible with surprising outliers or simply just luck, and it has changed the whole future of medicine. If Fleming did not put his Staphylococcus plate on the lab bench, if the fungal laboratory was not right downstairs of Fleming's lab, if the season was not the summer with the perfect temperature and humidity that fungi prefer the most for growth, if the lab windows were not open, if the wind did not blow into Fleming's lab, none of these discoveries would have been possible. Nevertheless, Fleming was able to purely separate and culture *P. notatum* successfully and published the results of using fungal juice containing penicillin into an infected mouse in the year of 1929, through *British Journal of Experimental Pathology*. After Fleming's publication, many chemists in England tried purification and separation of penicillin, but all of them failed. In 1939, an Australian pharmacologist and pathologist at University of Oxford, Howard Florey (1898 – 1968), and an English biochemist, Ernst Chain (1906 – 1979), successfully separated and purified the penicillin from culture broth of *P. notatum* and proved the efficacy of penicillin. As a reference, Howard Florey is one of the most respected scientist in Australia. Not too long after Florey and Chain's profound findings, they proved the significant and strong efficacy of penicillin in mouse and human diseases. Through that, the scientists in the world successfully found and separated a new penicillin strain called *Penicillium chrysogenum*, with the help of USDA (US Department of Agriculture) and several different pharmaceutical companies. With the successful mass production of *P. chrysogenum*, many lives were saved during World War II. With the recognition of its importance and success, Fleming, Florey, and Chain shared the Nobel Prize for Physiology

Figure 2-3 Alexander Fleming in Scottish banknote

or Medicine. Through a series of penicillin development, technologies on strain development, mass production, and separation and purification of final product contributed immensely not only in the development of antibiotics, anticancer agents, and antiviral agents but also in developing new functional agents. As shown, penicillin-related researches are known to be one of the most incredible stories in the science history. As Fleming mentioned, many of the scientific accomplishments are found accidentally. These types of findings are called as 'serendipity'. Figure 2-3 shows that Fleming copied on Scotland paper money.

As of right now, there are massive amounts of uncountable viruses, microorganisms and various diseases that are caused by them. Many types of antibiotics, anticancer, and antiviral agents are being developed to fight against these viruses and microorganisms, and these medicines ultimately come from bioresources that exist on earth. Especially, appropriate protein drugs are emerging as scientists are finding more about mechanism of human metabolism. The protein drugs are normally produced by the recombinant strains that was developed through genetic recombination of an organism. Ever since FDA (Food and Drug Administration) allowed distribution of insulin in 1982, many protein drugs have been produced and sold, and more new and upgraded ones are expected to be developed continuously with the use of bioresources. (Figure 2-4)

Environmental problems should be seriously considered in order to keep the ecosystem healthy and protected. The human society has already been damaged so much socially and economically, and the environmental pollution is still happening in the current state throughout the whole world. Through this situation, no one can expect confidently what the future holds on the Earth. The three E's – Energy, Environment, and Economy – directly and indirectly affect each other. As of right now, about 7.4 billion human population are living on this earth. Let's simply calculate this. If the human population increases twice as much, the size of the world's economy will also increase twice as much. This increase will then increase energy consumption which can accelerate the global warming and environmental pollution.

In 2010, the Nam-Seoul Museum of Art displayed the artworks of fifteen Asian artists in the theme of 'Ancient Future'. The exhibition topic

Figure 2-4 Application of bioresources to various pharmaceutical areas

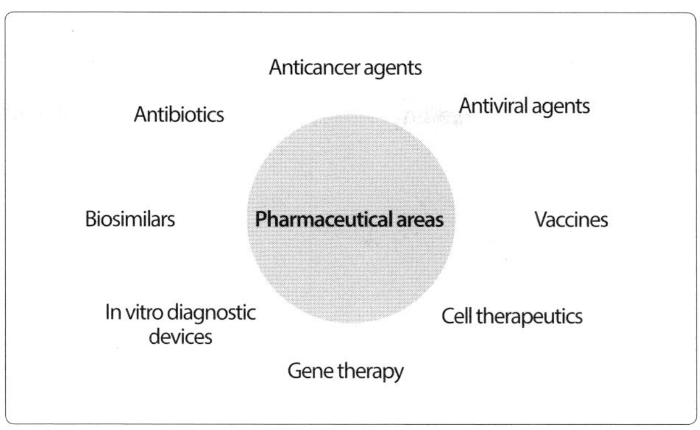

was organized to carefully think about the nature and civilization. In December 16th, 2010, there was an article written under the art category of the Korea Chosun Daily, which pierced my heart. The quote said, "In the last century, human beings have been having too much fun eating, drinking, using, and throwing away things. By having too much fun, did human beings irrevocably disrupt the nature of earth where they have rooted into?". I hope this quote also helps the readers to think thoroughly about the importance of keeping our earth healthy.

Environmental pollution is being easily observed in the atmosphere, water, and soils right now. Mechanical and chemical treatments were mainly used in the past to remove or recover these pollutions. However, in present, the use of biological treatments has been highlighted as pollution treatments. Here is the example. The soils around the specific chemical factory might be polluted due to TNT and other chemicals, and it will take a long time for the soil to be recovered back to the normal state. That is why many studies on biological treatments are actively under research to find more efficient way to treat pollution in a short time. One of the efficient way is to study the chemically polluted soils and find microorganisms that use these chemicals as nutrients. This means that the microorganisms are surviving in the soil by using these chemicals as nutrients. This shows how important microorganisms are as bioresources, and shows the possible use of these microorganisms to recover chemically polluted soils. This

Figure 2-5 Application of bioresources to environmental areas

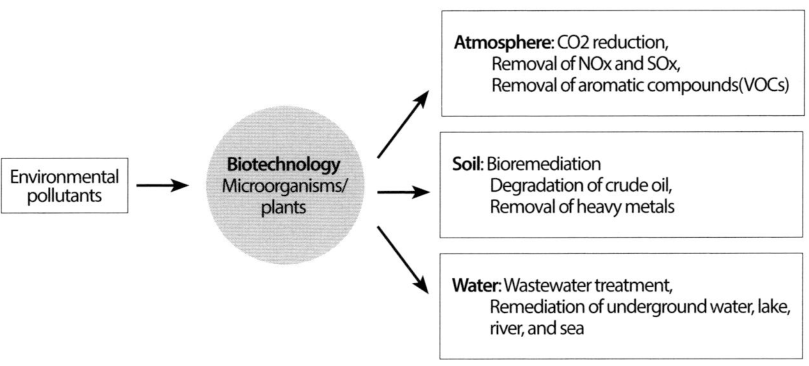

is not the only thing that the microorganisms can do. Microorganisms can also decompose dioxin, which is a strong carcinogen found in the atmosphere. In addition, some microorganisms are made as microbial agents and are used to decompose crude oil and fuel that are released from ships in rivers or oceans. In the recent years, the water pollution has been getting worse due to green algae and red algae. Increased water pollution is also occurred by heavy metals, nitrogen and phosphorus salts, various organic materials, and crude oil. An example of heavy metal pollution is a Minamata disease that occurred in Japan. In the 1950s, many people died and high rate of deformed infants were born due to mercury poisoning. This incident took place due to waste water from a chemical plant in a small fishing village. Not only that, the water can be polluted by cadmium that was produced within an industrial waste. Water polluted with cadmium can cause cadmium poisoning, which can cause Itai-itai disease that causes weakened and brittle bones in humans. In 2008, phenol was detected in the Nakdong River in Korea, so the domestic water intake was prohibited for a while. Throughout many of these pollution problems, the use of various microbial agents is a great source to solve the problem. Biotechnology field is developing as the desire to solve the environmental pollution problems is growing, and this growing area is called the environmental biotechnology (Figure 2-5).

Use of energy is also going through the similar problem. About 97%

Figure 2-6 Fluctuation of crude oil price(Dubai crude)

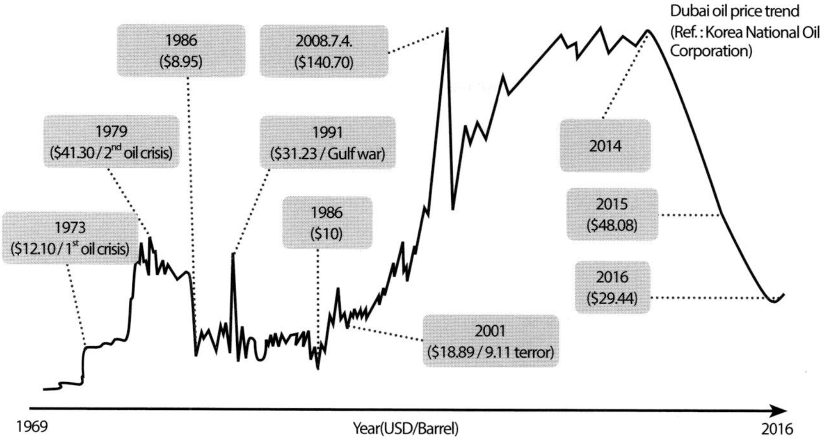

of the energy that is used in South Korea is being imported from different countries. As shown in the first oil shock in 1973, the crude oil price had hiked up from $2 to $10 per barrel due to crude oil embargo on the export by Middle Eastern oil-producing countries. In the 1980s during the second oil shock, the oil price had gone up to $42.25 per barrel. The oil price kept going up to over $100 per barrel by the year of 2007, and had settled down to about $110 per barrel in June 2014 with the basis of Dubai oil price. However, the graph of oil price started going down. By December of 2014, the price had gone down to $50 per barrel, then to $30 per barrel by the early 2016. This fluttered trend is due to political and economic relationships based on energy between the oil-producing countries. The shale gas that is developed and sold in the United States also seems to play a big role in this. As a non-oil-producing country, the only thing that can do is to wait until everything is settled down. However, the petrochemical industry and industries in general are hitting the bottom as the oil price change is fluctuating massively due to global political and economic situations (Figure 2-6).

With the witness of the volatility of oil prices, we have experienced how much energy can affect in overall social economy of a country. Therefore, many countries are emphasizing on developing renewable energy such as

Figure 2-7 Various kinds of biomass

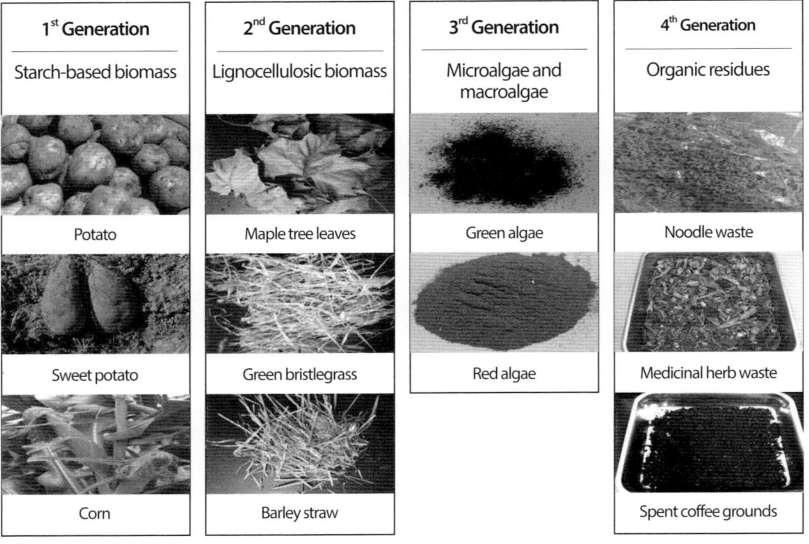

solar heat, solar power, wind power, geothermal power, tidal power, and bioenergy, in order to reduce the dependence on imported energy. Among those, bioenergy is type of energy that can be produced from biological resources. The important bioresources that are used in production of bioenergy are called biomass. Biomass contains all organic compounds originated from animals, plants, and microorganisms. In addition, biomass mostly refers to the agricultural, forestry, urban, and industrial wastes. Also, many research institutes in the world are focusing on development of energy crops. Biomass can be divided into different generations. The first generation of biomass is various grains, but the price of grains is now too high to be used as energy. The second generation of biomass is lignocellulosic biomass, which refers to trees and agricultural wastes. The third generation is microalgae and macroalgae that are widely distributed in the ocean, and the fourth generation contains all the organic wastes. By using these various types of biomass as raw materials, they can be converted into bioenergy with the help of biotechnology (Figure 2-7).

As of right now, bioethanol, biodiesel, biogas, or biohydrogen are used as different types of bioenergy (Figure 2-8). As an example, this is how bioethanol is produced. Glucose, a simple sugar made with six carbons, is formed after breaking down the pretreated lignocellulosic materials such as wood, straw, fallen leaves, or fruit peels with the enzyme called cellulase. The glucose can be converted into ethanol after the fermentation process, which is the reaction that occurs when glucose is consumed by yeast. The final product of ethanol can be concentrated and be used as fuel for transportation. As the importance of lignocellulosic materials is shown above, the amount of these lignocellulosic materials should be used with careful consideration of the right seasons and weathers.

Figure 2-8 Various kinds of bioenergy

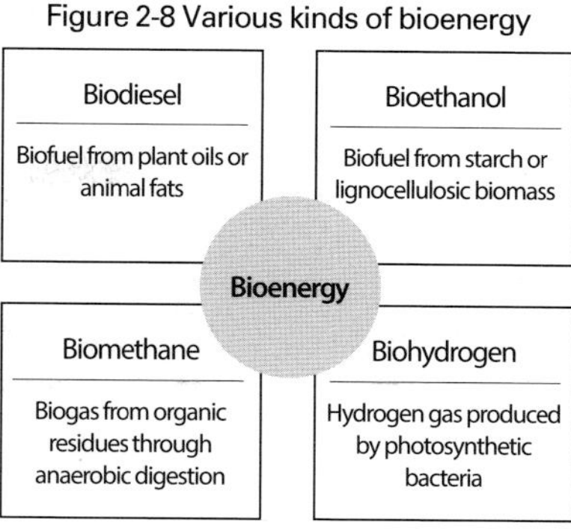

Biodiesel is a fuel material that is very similar to diesel. Biodiesel can be produced through two different methods: the chemical method, which treats oil or waste oil from industries with chemical catalysts, and the enzymatic method, which uses lipase, an oil-degrading enzyme. The most widely used oil products throughout the world are mainly vegetable oils, such as palm oil, soybean oil, canola oil and waste oil, but animal oil products are also used. A large amount of glycerol is produced as a by-product during the production of biodiesel, and thankfully, they are used as one of the main ingredients in various fields including cosmetics production.

When anaerobic fermentation is used with various microorganisms using organic wastes such as industrial wastes, agricultural wastes, animal wastes, and municipal wastes, organic acids are produced. The organic acids are then converted into methane gas, which is called as biogas. Biogas can be used as energy in the homes like natural gases.

Biohydrogen can be produced with either chemical or biological procedures. With the development of biotechnology, many researches in regards to using diverse microorganisms including photosynthetic bacteria that can produce hydrogen are becoming more popular. The importance of hydrogen is being emphasized due to its availability of use in variety

Figure 2-9 Concept of biorefinery

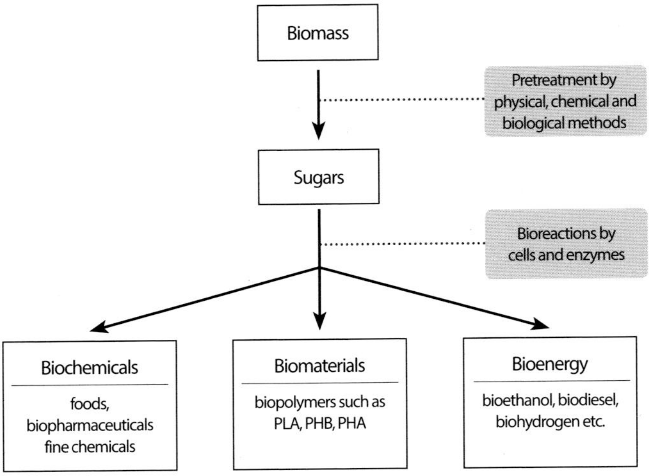

of future energy sources because of its characteristics and ability to react with oxygen to make water (H_2O) and going back to its previous stage through bond breakage.

All of these raw materials and final products of bioenergy are part of bioresources, and many of the research institutions and industries are putting in much effort for industrialization. For the nations with limited raw materials, establishing a process of bioenergy production with their own technology and using the different organic waste materials may be a good idea.

Recently, some industries use the concept of Biorefinery. Biofuels, chemical intermediates, and the end products are produced through obtaining sugars from biomass. Also, synthetic gases are produced through a direct thermochemical process of biomass to be used as energy. As shown above, industries are developing different processes of producing biofuels, chemical intermediates, and the final products (Figure 2-9).

Food products contain all edible food except any ingredients that are used in generic drugs, and can be classified according to the method of production, raw materials, ingredients and the usage. Out of many of these food products, traditional foods, functional foods, and GMOs are mainly focused in the biotechnology field (Figure 2-10). Some common Korean

Figure 2-10 Classification of food

General food	Health functional food	Functional food
All the food containing functional and health functional foods	These foods have a useful function for human body, maintaining or improving a health by activating a physiological function	These foods generally have a function of biological regulation, having a broader meaning than health functional foods

traditional foods that can be easily found in any places are Kimchi, soy sauce, fermented bean paste, fermented chilli paste, and cheonggukjang etc.. Korean Kimchi has become one of the world's healthiest food in 2001 as CODEX ALIMENTARIUS selected Kimchi as an international standard. In 2010, World Institute of Kimchi (www.wikim.re.kr) was established in Gwangju, Korea. There are over two hundred kinds of Korean Kimchi, and hundreds of bacteria play a role in proper fermentation with good taste. In general, Kimchi is vegetables submerged in salted water. The primitive Kimchi might have been the simple pickled-cabbage just for longer storage period during the winter. As time went by, many different types of sauces and vegetables were introduced to Kimchi fermentation. Seeing the time when multiple vegetables started to be imported into Korea, the ordinary cabbage Kimchi was thought to be emerged in the 1700s. Although the modern way for Kimchi fermentation is to simply put it in a topper wear made for a special Kimchi fermenting refrigerator, the ancestors used to put Kimchi in a big, sealed pots, and bury it under the ground to minimize the temperature change. This Kimchi pot was used as a bioreactor for appropriate fermentation. During this process, Kimchi produces various types of bacteria that produce lactic acid, which is known to have anti-bacterial ability against toxic germs. This process was also found to be healthy for human gut microbiota. In the recent years, many scientific papers were published in regards to physiological active substances and functional substances produced during fermentation having anticancer and antibacterial effects (Figure 2-11).

An example of a functional food is novel sweeteners. As human body cannot degrade novel sweetners, it cannot be converted as calories, which is good for obesity prevention. Not only that, growth of Bifidobacteria will improve the environment for intestines when the novel sweetener reaches

Figure 2-11 Fermentation and function of Kimchi

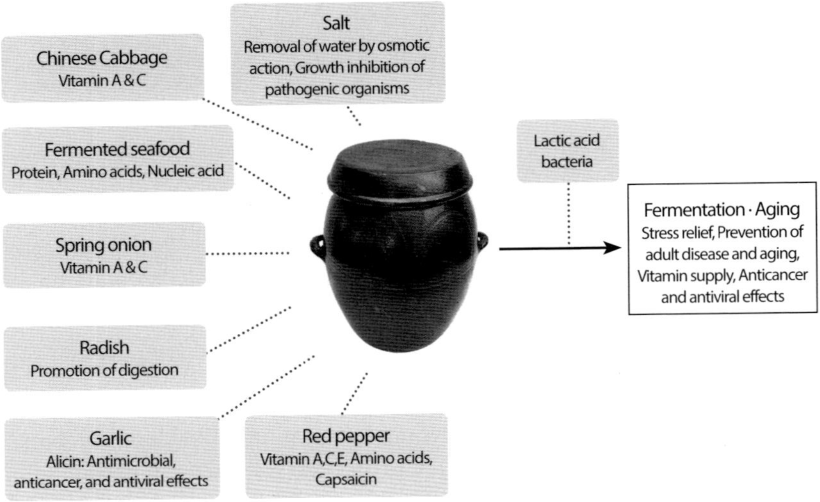

to the intestines. These sweeteners are currently being used in homes instead of normal simple sugars. For alcoholic beverages, oligosaccharides, aspartame, and stevioside are used for a sweet taste. Besides, various sweetners were used for gum, candies, ice cream, sodas, bread, snacks, and dairy. There are other functional foods such as astaxanthin, a strong antioxidant, or coenzyme Q10, β-carotene, and lycopene etc.. Figure 2-12 shows the function of the functional foods.

Another important part of food is GMOs. Since GMOs were discussed in depth in the earlier chaper, it will be skipped here.

Thus far, bioresources that are being used in many different fields were introduced for better understanding. Though there are so much more in the use of bioresources, it is important to think carefully first in the use of bioresources with social, cultural, political, and economical perspectives.

Starting from October 6, 2014, the 12th Convention on Biological Diversity (CBD) was held for two weeks with the topic of 'Biodiversity for Sustainable Development'. CBD was selected in 1992, during the UNCED that was held in Rio, Brazil. Then, it came into effect on December 29, 1993. South Korea applied for it on October 3, 1994, and took effect on January 1, 1995. The current status from June 2014, about 194 countries

Figure 2-12 Various kinds of functional food and their functions

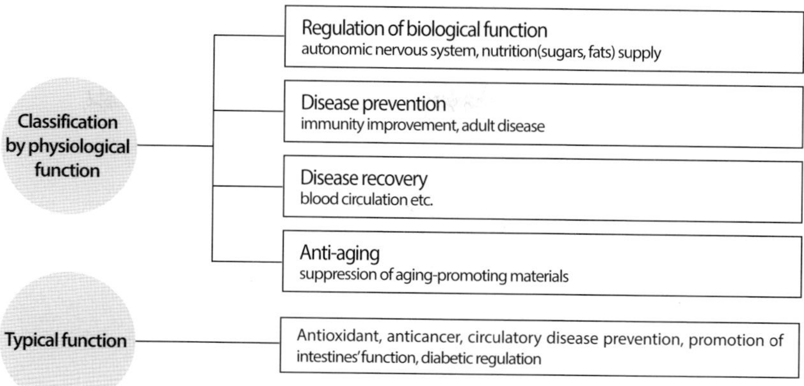

are under this convention. The purpose of CBD is for conservation of biodiversity, sustainability of bioresources usage, and equal share of benefits obtained from the utilization of biological genetic resource.

The ‹Nagoya Protocol› was officially activated on October 12th, after the 12th Convention on Biological Diversity. As this protocol has become effective, various industries are thought to be highly impacted by it. For the countries that are importing bioresources from other bioresource-rich countries, it will be important to find secondary solutions and to consistently discover and register the native bioresources in land.

The countries that have the richest bioresources are called Megadiversity 7 (M7). These countries are Australia, Democratic Republic of the Congo, Madagascar, Columbia, Mexico, Brazil, and Indonesia. They own over 50% of the whole world's bioresources.

As an example, the Q Millennium Seed Bank that was found from 1992 Rio Conference, has been playing a critical role in reviving plant species that are at a risk of extinction.

The importance of conservation and possession of biodiversity of a country is thought to be emphasized more in the near future due to the constant growth and needs of bioindustry. Therefore, each nation should carefully draw and plan the future blueprint of the bioindustry.

Chapter 3

Organisms as a Perspective of a Small Factory

As emphasized throughout the previous chapter, bioresources from microorganisms, plants, and animals are extremely important for the industry. Anton van Leeuwenhoek (1632 – 1723) was a Dutch scientist and the first scientist to find small molecules and organisms living in a pond that cannot be seen with the naked human eyes by using a microscope. These small and almost invisible organisms were named as animalcules (Figure 3-1). The diameter of a smallest bacterium is only 0.1-0.2㎛, and the diameter of an *Escherichia coli* is normally 0.5㎛. The biggest cell considered is the cells of a bird or a mammal egg. Especially, the whole chick egg yolk is considered as a single cell with the diameter of 50 mm.

The concept of a cell was named by an English scientist, Robert Hooke (1635 – 1703), when his observance of a cork oak bark reminded him of the small rooms clustered in a monastery. Then, Robert Brown (1773 – 1858), an English botanist, discovered a nucleus of a cell, and a German botanist Matthias Schleiden (1804 – 1881) announced the plant cell theory. As a German physiologist, Theodor Schwann (1810 – 1882) had

Figure 3-1. Various microscopes

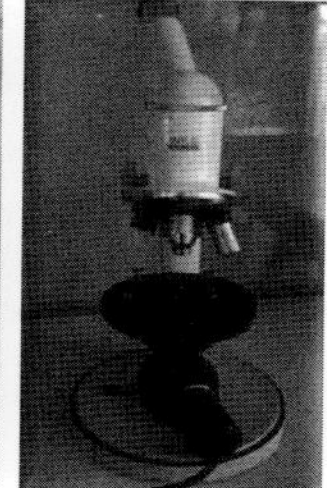

also announced the animal cell theory, many scholars have made a great contribution to establish today's cell theory. Although different types of cells are composed of slightly different types of organs, they are all built with atoms and molecules, and share common components that have same functions. The examples are plasma membrane, cytoplasm, organelles, and cell nucleus. Cytoplasm is surrounded by the plasma membrane, which is the outermost layer of a cell. In all cells, the plasma membrane surrounds the cytoplasm and protects against foreign organisms and substances. One of the most important organelle in a cell is mitochondria. Mitochondria is very crucial for cellular metabolism by producing an energy source called Adenosine Triphosphate (ATP). Nucleus of a cell has the overall control of the cellular activity, and contains all genes of a cell. Since the details of structures and roles of these cell organelles are well-explained in many textbooks, we won't go further in here.

In general, there are some important definitions to be considered as a living organism. Cells are the fundamental building blocks of living organisms. Each and every cell goes under mitosis for chromosome replication and separation. All cells are protected by the plasma membrane. Cells can also transform and use energy to preserve genetic information from a mother cell, and transfer the exact same genetic information to a daughter cell during meiosis. All these definitions rely on the Cell Theory where all cells come from cells. Having basis on genetic information, cells can synthesize various types of proteins, making necessary substances and energy via many different types of metabolic pathways, and has a signaling system for external stimuli such as temperature and pH change.

On the other hand, virus used to be defined as an inanimate object in the past, but now has changed to a parasitic, non-cellular infectious agent. Because virus can only survive and metabolize inside of a host cell, it is considered as a parasitic agent. The word 'virus' originated from a Latin word that describes toxin. Virus can be classified into DNA virus and RNA virus, and each virus can be divided into single stranded or double stranded. Viral genetic information is protected by a protein called capsid. The virus that can infect and replicate within a bacterium is called bacteriophage. There are two mechanisms of common viruses. Normally, a virus replicates its own genetic material by using the host cell's organelles. Most of the time, the host cell simply gets damaged. In other case, it may end up having a disease due to more serious destruction during this process. However, some host cells can just act as a mediator without the disease break-out even if they are infected by a virus.

Figure 3.2. Potential applications of virus

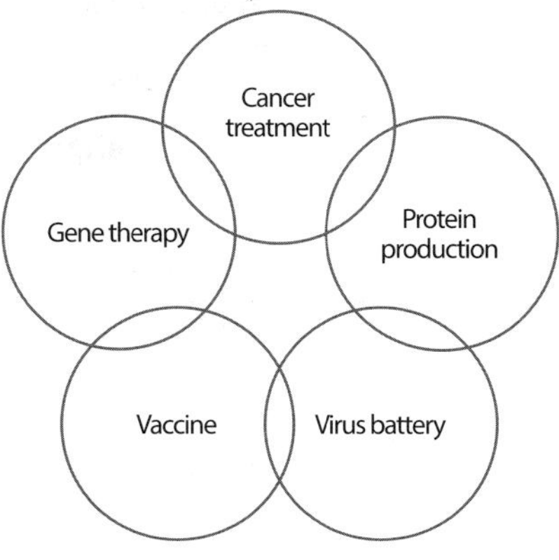

In the recent years, many ongoing studies have been focusing on the application of functions or characteristics of viruses such as creating useful substances or energy by using substances pulled out from a virus or virus itself as a possible vector for gene therapies and cancer treatments. Some significant studies focused on mass production of plant-based proteins by using plant-derived virus, and development of nanogenerator by synthesizing barium titanate (BaTiO3) nanocrystals from genetically modified M13 bacteriophage (Figure 3-2). As of right now, studies on RNA viruses such as retrovirus and adenovirus, and their application as gene therapy vectors in cancer treatments are actively under research.

Humans are generated from one embryo, which is considered as a single cell. As the embryo goes through multiple cellular divisions, about 3 trillion cells are made by the time the baby is born. As the baby grows to be an adult, about 50-100 trillion cells are present. As humans live, cell division continues, and dead cells are continually falling onto the floor as 'dead skin'. All cells in a human contain exactly same DNA in the nuclei. However, these cells and their DNA are slightly different in each human. Red blood cells do not have nuclei. One interesting fact is that a mitochondria, an organelle in a cell, contain its own mitochondrial DNA (mDNA). mDNA is useful when studying phylogenetics, ecology, and anthropology because of its trait in descending only from the maternal side. Therefore, mDNA is also called an Eve gene. There is also an Adam gene: Y chromosomes in men. Y chromosomes are mainly used to track the

Figure 3-3. The scientific name(Genus and Species) and coined word according to the evolution of mankind

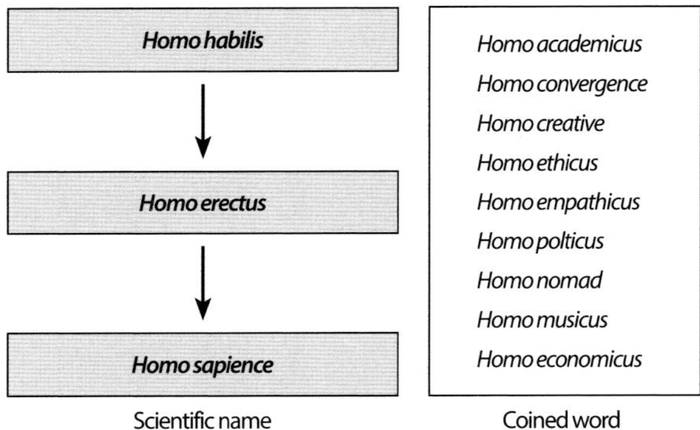

movement of past different races. Dr. Wook Kim in Dankook University tracked the Korean race by using Y chromosomes and showed that Koreans have mixed races of 70-80% northern race, 20-30% southern, and other smaller groups.

A plant organelle, chlorophyll, also contains its own DNA and mRNA. Many bacteria do not only have their own DNA, but also have circular DNA called plasmid.

Any living organisms are organized by kingdom, phyla, classes, orders, family, genus, and species. A Swedish scientist Carl von Linné (1707 – 1778) first introduced to use binominal nomenclatures in describing different organisms by using genus name and species name. For example, *Escherichia coli (E. coli)* is written with the first 'E' capitalized, and this binominal nomenclature is expressed with either italics or underlined. The modern humans are called *Homo sapiens* sapiens. The last sapiens shows the subspecies, and *Homo sapiens* means 'a person with wisdom'. Many fun words have arisen lately to describe the modern humanism. As convergence technology and creative technology have developed, *Homo convergence* or *Homo creative* were made. As social network service (SNS) has developed, *Homo digitalis* was newly expressed to call the human beings (Figure 3-3).

Robert Whittaker (1920 – 1980), an American ecologist, divided the living organisms into five different groups: Animalia, Plantae, Fungi,

Figure 3-4. Various kinds of archaebacteria

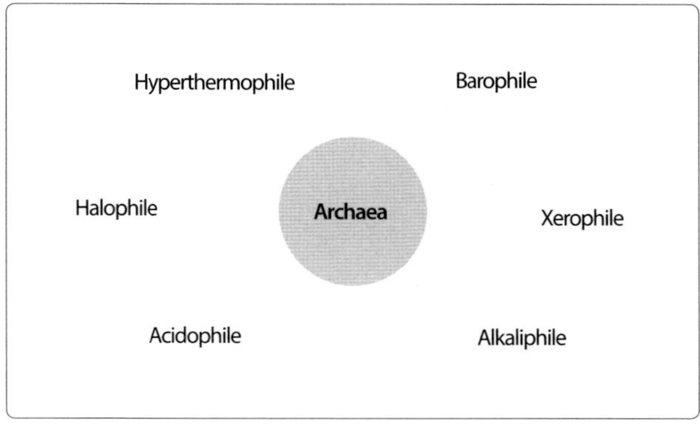

Protista, and Monera. There are many other ways to divide organisms into different groups. Monera can be further divided into eubacteria and archaea. Most of the bacteria in the Archaea group are very ancient bacteria. Because most of them reside in extreme environments, they are also called extremophiles. Due to the environmental difference, these extremophiles normally have quite differently composed cell walls, cell membranes, and nucleotide sequences. Another interesting fact is that they also have different metabolism. Good examples of these archaea are thermoacidophiles that can live in high temperature and low pH or halobacteria that can live in the areas with high saline concentration. Archaea are known to be very useful in the industries, therefore, are highly interested by the researchers and are actively under research (Figure 3-4).

In general, cells can be considered in two different types – prokaryotic cells and eukaryotic cells. A significance of a prokaryotic cell is that it does not have a nuclear membrane, therefore, the nucleus is not separated from other cellular parts. Also, a prokaryotic cell has the characteristic of monosomy, and is made of few cell organelles. Single cell types such as bacteria and archaea have the characteristics as above. A eukaryotic cell has rather more complex organization. A nucleus is surrounded by a nuclear membrane and has high numbers of chromosomes. Within the cell, various types of organelles exist. All eukaryotic cells such as fungus (yeast, mold, mushroom), protist, plant cells, animal cells have multicel-

Figure 3-5. Prokaryotic and eukaryotic cells

	Prokaryotic Cell	Eukaryotic Cell
Nuclear membrane	X	O
Chromosome	Single	Multiple
Nucleolus	X	O
Mitochondria	X	O
Chloroplast	X	O
Endoplasmic reticulum	X	O
Golgi complex	X	O
Cell wall	O (Peptidoglycan)	O (No cell wall in animal cells)

lular morphology. One interesting fact is that unlike a normal eukaryotic cell, an animal cell does not have a cell wall (Figure 3-5).

Also, a Danish bacteriologist Hans Christian Gram (1853 – 1938), developed the Gram staining technique with the basis of different composition of cell wall of bacteria. With the use of the Gram staining, Gram differentiated bacteria into groups of Gram positive and Gram negative. Gram positive bacteria such as *Bacillus subtilis* normally do not consist of outer cell membrane, but have a thick, solid, multi-layered peptidoglycan. Gram negative bacteria like *Escherichia coli* have both outer and inner cell membrane. The inner cell membrane is made of multiple proteins, lipids, and carbohydrate layers, and the outer membrane is made of a thin peptidoglycan.

Having understood the basics of cells for their use in the industrial field, let's consider that a cell is a small factory. In order for a small factory to produce the final product, many small processes have to work through consistency. A cell works the same way as the small factory. The only difference is that the factory has to change its process-line program depending on its use, but a cell adapts and changes its own metabolism proper to the outer environment. As the modern science has improved, the scientists have achieved to discover the biological metabolic processes of a cell and to control the parts of a cell's metabolic pathway with physical, chemical

Figure 3-6. Example of metabolic pathway

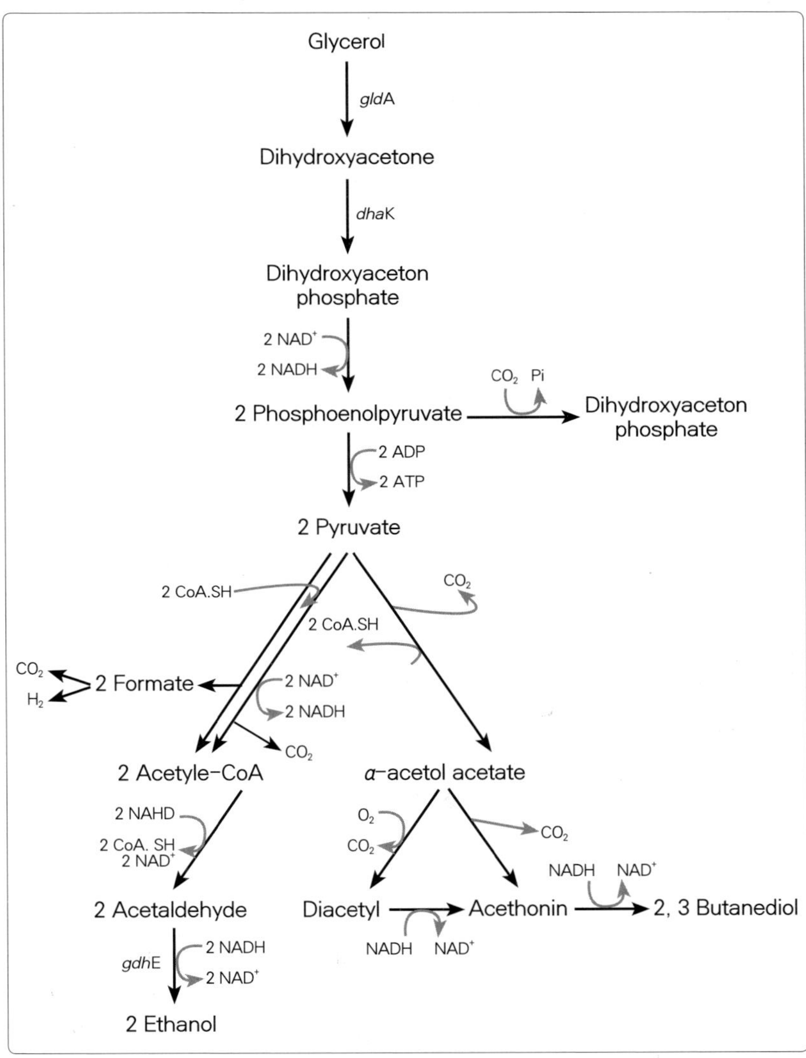

and biological stimuli. This field is called metabolic engineering, and it is a technology that can artificially control the cellular metabolic pathways for mass-production of biotechnological products. Together, the field of synthetic biology is massively growing (Figure 3-6).

For example, Craig Venter (1946 -), the founder of Celera Genomics

Figure 3-7.
Fermentation process

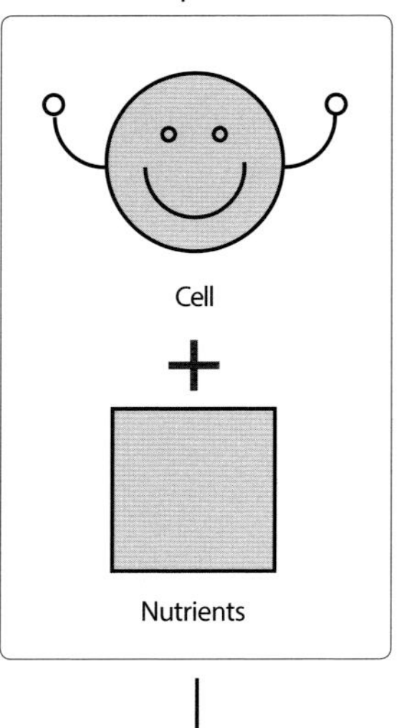

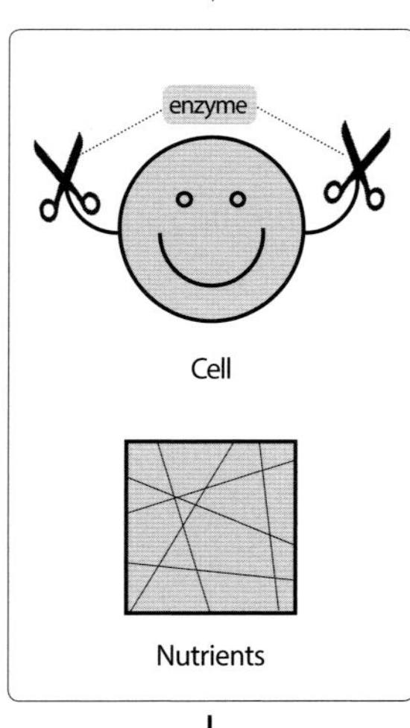

Primary and secondary metabolites

who leads the Human Genome Project, published an article titled "Creation of a bacterial cell controlled by a chemically synthesized genome" on "Science" journal in 2010. This article introduced an idea of making an artificial microorganism by deleting the genome of *Mycoplasma capricolum* and inserting a genome of synthesized *Mycoplasma mycoides* DNA which was constructed in yeast. Also, the March 2016 article on the "Science" titled "Design and synthesis of a minimal bacterial genome" published that an artificial organism called JCVI-syn3.0 was made. This new organism has the size of 1/6th of base pairs of normal bacteria, around 531 kbp and 473 genes. These kinds of new technologies are thought to play an important role as a customized 'cell factory' to produce multiple products in the future.

Although a cell is organized in many different ways, it generally absorbs nutrition to make energy and various metabolites through different metabolic pathways in order to use immediately or save up for the future. A cell itself is made up of basic elements such as carbon, hydrogen, oxygen, nitrogen, and minerals. Therefore, in order for a cell to produce metabolites and energy through metabolic pathways, it is necessary to supply various nutrients such as carbon, nitrogen, oxygen, minerals, vitamins. This metabolism can be divided into anabolism or catabolism. Anabolism synthesizes simple atoms into more complex biomolecules, and catabolism breaks down these biomolecules into simple atoms. During these processes, energy is either yielded or released. On earth, many organisms are mainly based on carbon, hydrogen, oxygen, nitrogen, phosphorus, and sulfur. In 2010, National Aeronautics and Space Administration (NASA) found a bacterium that intakes toxic arsenic for growth, at the Mono Lake in California. The discovery of the bacterium that absorbs arsenic as a nutrient can be a vital evidence for the possible living organisms outside of the earth.

As explained above, cellular pathways have been applied in many different areas such as food, pharmaceutical, agricultural, or environmental fields. Especially, fermentation has been used since long ago. In the beginning, the idea of it was only to produce alcoholic beverages by yeast fermentation. However, the modern industrial fermentation includes all processes of manufacturing bioproducts by mass culture of microbial cells. The biochemical meaning of fermentation is the process of energy production during decomposition of organic compounds. The word fermentation originated from Latin, which carbon dioxide diffusion resembled water boiling during ethanol fermentation of yeast. Though, production of useful components from animal or plant cells are not considered as fermentation. Instead, they are called animal or plant cell culture (Figure 3-7).

Mass culture of cells indicates that a large amount of products can be produced at once as the number of cells increases because of the increased 'cell factories'. Also, the process includes all units of procedures in order to get to the ultimate purpose. Increased cell factories do not always produce large quantities of desired products. The environment has to be at the happy place for the cell factories to produce these products, such as right levels of temperature, pH, oxygen concentration, types and concentration

Figure 3-8. Various factors affecting cell factory

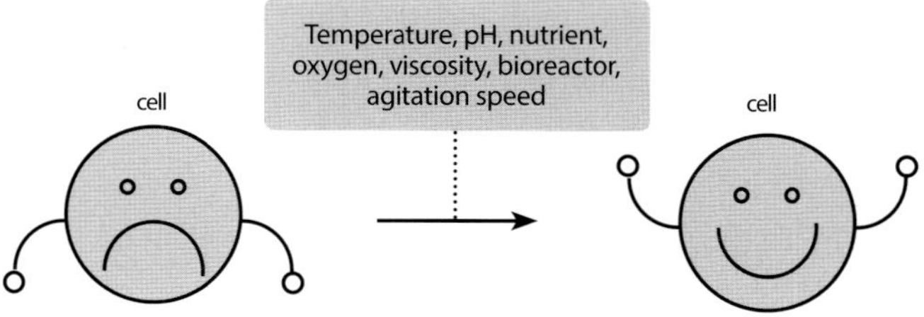

Figure 3-9. Applications of yeast

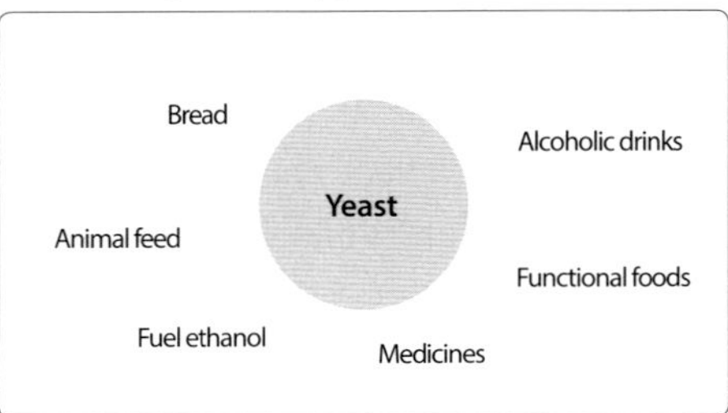

of nutrients, and bioreactors. Therefore, in order to produce desired products properly from specific cell types, understanding the basis and characteristics of the used cell type is crucial (Figure 3-8).

Through the cellular mass-culture, we can generally collect certain useful components such as the cell itself, enzymes, metabolites, or bio-transformation. To explain more about each component, we will start with yeast. Yeast is known as the product of a cell itself. Industrial yeasts include ethanol fermenting yeasts and bread fermenting yeasts. These types of yeasts are distributed and sold by the markets, therefore, are the product as they are. In this case, yeast cells are mass-produced, then manufactured as either wet or dried, packaged, and distributed to the markets. Wet yeast can be contaminated easier by other microorganisms, therefore, has shorter life span and should be refrigerated. On the other hand, dried yeast with under 5% of hydration has longer expiration date, so they can be exported throughout the nations (Figure 3-9).

Enzymes can be divided into three different groups: industrial enzymes, diagnostic enzymes, therapeutic enzymes. Industrial enzymes include carbohydrases, proteases, lipases, cellulases and pectinases. Normally, industrial enzymes can be useful in many different fields with even lower purity. Due to this, they can be partially purified and the cost is lower. Diagnostic enzymes include oxidases and reductases, and are used in bio-chips such as protein chips and biosensors. These diagonostic enzymes are mostly applied in finding out specific diseases or materials. Even though the purity is not 100%, it is maintained to 95-98% by a series of partial purification. Though, therapeutic enzymes need to maintain the purity to about 100%. Digestive enzymes, anti-inflammatory enzymes, and throm-

Figure 3-10. Examples of industrial, diagnostic, and therapeutic enzymes

Industrial enzymes	Diagnostic enzymes	Therapeutic enzymes
Amylase	Glucose oxidase	Streptokinase
Protease	Alcohol oxidase	Urokinase
Lipase	Cholesterol dehydrogenase	Adenosine deaminase
Cellulase		Digestive enzymes

bolytic enzymes are such therapeutic enzymes.

If impure therapeutic enzymes are injected into human bodies, a patient can show the side effects such as allergic reaction or severe toxification reactions. Since it takes time and money to go through multiple steps of purification, therapeutic enzymes are the most expensive out of all three types of enzymes. Also, most of various recombinant enzymes are also included in this category (Figure 3-10).

There are two types of metabolites. Primary metabolites that are fundamental in cell growth, and secondary metabolites that are not directly associated with cell growth. Primary metabolites include amino acid, nucleic acid, protein, and organic acids that are crucial for cell growth. Secondary metabolites are alkaloids, flavors, and antibiotics that play a role in cell protection mechanism mostly. Primary and secondary metabolites can be produced in different ways. Primary metabolites are produced as the cell grows, or a cell can convert to secondary metabolites from the primary metabolites that were already produced, or a cell also produces secondary metabolites from the nutrients directly that were nothing to do with primary metabolites. Another interesting fact is that modern biotechnology can produce a brand-new substance by modifying the chemical or biological composition of the metabolites that were mentioned above. Good example of these new substances are steroids and antibiotics.

As mentioned above, nutritional compositions and concentration, and culture conditions should be carefully controlled in order to keep the environment optimal to produce these useful products from the cell factory. Although technology is the most important part in this, understanding the basics of historic, cultural, economic, social, and political aspects of the related science, engineering, and industry is very important.

Let's say we want to produce high value-added products. To do that, market research on the specific economy should be prioritized. If the product is considered to have the worth for mass production as the market research goes, the basic data on product manufacturing should be carefully collected. Nowadays, collecting data are much easier than before, thanks to the tool called big data. Big data has been a great help in many different fields in collecting large amounts of data that is needed. Big data made into top ten highlighted noble scientific technologies a couple of years ago by Davos Forum. In the list, synthetic biology and metabolic engineering that were mentioned above, were included. In a technical aspect, basic data gathering starts in a laboratory because choosing the right cell type is the most important factor during the starting point. It is important to know if the cell can produce the desired product, and if the cell can go through mass production. The possibility of mass production of a cell depends on the cell activity, yield, and productivity. These factors are very important because they have direct impact on the economy. Activity of a cell explains if a cell can produce consistent amount of products, and this is why obtaining a healthy cell is crucial in the economy field. When bio-based companies develop a highly active and healthy cell, these type of cells are mass cultured for inoculum, freeze-dried or placed into small containers with glycerol, kept inside the deep freezer, and taken out one by one when used (Figure 3-11).

Yield shows the ratio of all released products when a cell consumed a certain amount of nutrients, or the ratio of its product and a constant cell concentration. Productivity is a kind of performance on production of bioproduct in a certain period of time. These two are very important in evaluating the efficiency of a cell factory. Since yield and productivity have a close relationship with economical efficiency, continuous analysis should be done.

Let's assume that we have found the right cell to produce the desired product. We will probably provide the right type and amount of nutrients, enough oxygen supply, and set up the environment into the desired culture condition. All these stages should be built as pure culture. In a cell culture, there are aerobic culture and anaerobic culture. Aerobic organisms require

Figure 3-11. Preservation of seed

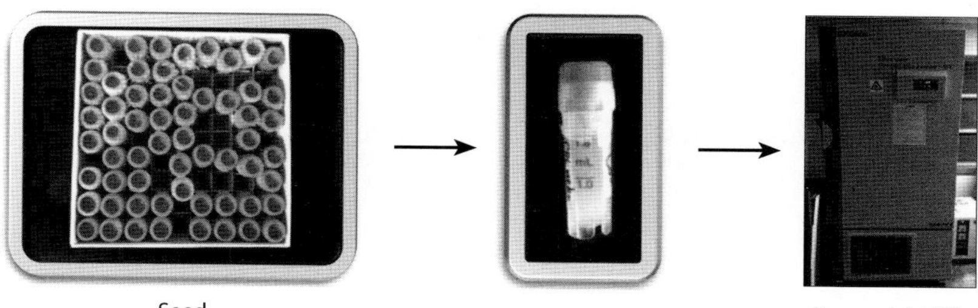

oxygen for growth, but anaerobic organisms grow only in the environments without oxygen. There are also facultative organisms that can grow in both aerobic and anaerobic conditions by controlling their own cell metabolism. Most of the organisms that are used in industry are aerobic organisms, but the ones that produce methane under the deep mantle or ocean are mostly anaerobic organisms. Yeast is a good example of a facultative organism.

During the aerobic culture, the culturing environment should be well-aerated for good oxygen supply. Due to the high cost of pure oxygen, air should be supplied constantly so that the air can be solubilized in water, and water-dissolved oxygen can be used by aerobic cells. In order for most of these cells to produce useful products, presence of oxygen is crucial. Also, pure culture means to culture only the type of cells that we know. If unknown cells are growing in the culture, it means the specific cell culture is contaminated by those unknown cells. Therefore, the environment must be kept clean and uncontaminated in any circumstances. If contamination occurred in mass culture in industries, it will be a massive loss for the company by not being able to make the desired product. For the purpose, every container, nutrients, or any tools that are used for the pure culture must be fully sterilized and the inoculum must be well-kept in the sterilized places.

As indicated above, nutrients include many different ingredients. The most important role of nutrients, however, is to help the cell to grow well and to produce useful products. In order for the nutrients to do the job, finding the right combination of ingredients for lower cost is crucial because this plays directly into the cost of the final product in the market.

Figure 3-12. Concept of Central Dogma

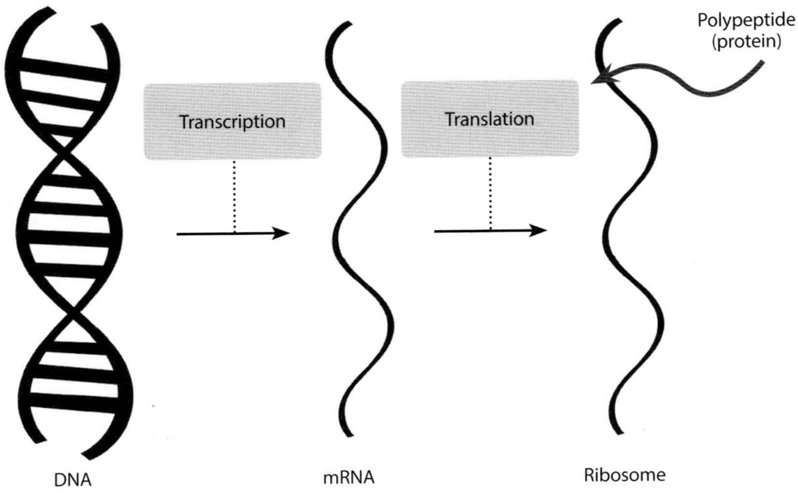

When a company is producing the cheap bulky products, 70-80% of the final cost comes from the raw materials of nutrients that was used to produce the product. Especially, carbon or nitrogen-contained nutrients are either soluble or insoluble, and this factor can make a big impact on the cell culture.

In cell culture conditions, the environment where a cell can grow well and yield products efficiently is very crucial. These conditions include culture temperature and pH. Bacteria can be divided into three different groups depending on their preferred growth temperature. The ones that grow under 20℃ are called psychrophiles, 20-50℃ mesophiles, and thermophiles if they prefer the temperature to be over 50℃. Most of the archaea that was mentioned above are extremophiles, which normally exist only in extreme environments.

The solution or slurry filled with nutrients with optimum culture temperature and pH is ready to accept the cells. Then, inoculum that was well-maintained needs to be transferred carefully into the slurry container through inoculation, and the environment should be completely free from contamination. Although the inoculum was introduced to the most desired environment, it takes time for the cells to adapt to the new environment.

This resembles a situation of a person being introduced to a new culture when traveling to a new country. Most of us need time to adjust to the new culture and the environment. As our experience is called a 'culture shock', cells go through a phase called lag phase. Every cell type goes through different length of lag phase. During the lag phase, cells go through something called Central Dogma, which underlines the production of multiple enzymes and proteins to digest and use the nutrients properly.

In here, Central Dogma explains the process of making a functional product, a protein, from the flow of genetic information, from DNA to RNA. The genetic information from DNA segment is copied, sent to RNA, RNA carries the information to the ribosomes to synthesize new proteins (Figure 3-12).

Once cells are well-adjusted by going through the lag phase, cells grow exponentially by ingesting abundant nutrients and oxygen under the optimum environment. During this time, important primary metabolites are made. This phase is called logarithmic phase or exponential phase. As many growing cells consume nutrients, the level of nutrients decrease, and this causes cells to slow down in growth. As the number of cells maintains, cells start to produce secondary metabolites that play a role in cell protection mechanism. This phase is called the stationary phase.

Figure 3-13. Cell growth curve in a batch culture

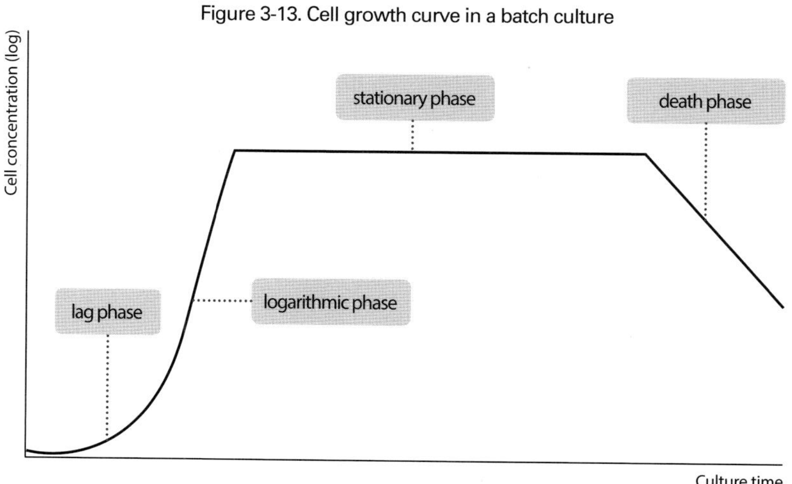

When all nutrients are used up, cells cannot maintain and the number of cells starts to decrease. This phase is called the decline phase or the death phase.

The stages above happen when there are limited amount of nutrients, and has the general growth curve of batch culture. In this type of environment, all cell types have different timing in reaching the maximum cell number. For example, *E. coli* takes about 5-6 hours, fungal cells take about a week, and plant cells take about two weeks to grow maximally. Thus, understanding the basis of cells and controlling the cell stages will help us to get the final product efficiently (Figure 3-13).

Now, let's see how a cell can properly ingest nutrients and oxygen. Most of the current aerobic cultures require oxygen, which is given as dissolved oxygen solubilized in water with soluble and insoluble nutrients. Water is very important in many aspects including the role above. Not only that, pH control is possible also because of water, and growing cells in the water can maintain their temperature consistently if the water temperature is well-maintained in the incubator or bioreactor due to the evenly distributed heat. Another important fact is the nutrients and oxygen that are dissolved within the water can be transferred into the cells. In other words, nutrients and oxygen can be well-distributed into the cell if the cell in the water can efficiently have contact with the nutrient and oxygen. Therefore, efficient and continuous mixing is crucial during this stage.

A cell goes through all the processes explained above, consume nutrients and oxygen, save up or release metabolites and carbon dioxide after going through complex metabolic pathways within the cell. The substances are released into water out of the cell. We can use many of these released metabolites in many ways. The final metabolite may be the cell itself or it can be released outside of the cell or accumulated within the cell. These metabolites can be carefully collected for the use after understanding their condition. Chapter 8 will go more deep into the recovery methods of metabolite.

As shown, cells that are selected and developed can act as excellent living factories. These amazing living factories greatly contribute to the welfare of mankind by producing useful substances. Enzymes are one of the most important substances that leads to successful production of various bioproducts. The next chapter will explain the basic concept and functions of the enzyme.

Chapter 4

Power of Enzymes Produced by Organisms

Enzymes that are produced by living organisms are generally organic catalyst called the biological catalyst or biocatalyst. As inorganic catalyst accelerates chemical reactions, biocatalyst is a type of protein that works the same way in living organisms. Biochemical reactions are also considered as chemical reaction, which either breaks or synthesizes organic materials. A lot of these reactions cannot occur without the help of these special enzymes. Although immense types of enzymes exist for all the cellular chemical reactions to happen, not all enzymes in cells have been identified yet in reality. However, higher creatures are thought to have more types of enzymes due to higher complexity of physiological reactions. So far, about three thousand enzymes have been identified, and around sixty of them were produced industrially.

Enzymes can also be classified into different generations. First generation enzymes are natural enzymes, second generation are immobilized enzymes, third generation are recombinant enzymes, and the fourth generation are artificial enzymes. The current industries mainly use the second and third generation enzymes, and further researches on the fourth generation enzymes are actively in progress. The enzyme-related technologies such as immobilization, DNA recombination, separation and purification, screening, and mutation have been constantly developing as the biotechnology fields have been popular, and are very important in many industries.

Before talking about enzymes, we will go more in depth about basic concept of protein.

The origin of the word 'protein' means 'of prime importance', and was first discovered by a Dutch chemist, Gerardus Mulder (1802 – 1880). Mulder found the chemical composition of protein, and named it as protein. Protein is a fairly large molecule and is chemically composed of carbon, nitrogen, hydrogen, oxygen, sulfur, phosphorus, and metals. Protein molecules change their tertiary structure depending on the combination of twenty amino acids with peptide bonds, and it changes the final function of each protein. Amino acids are easily dissolved in water, and are either acidic or basic depending on the structure.

Proteins are normally organized with pure amino acids. However, there are complex proteins that are composed of amino acids with organic or inorganic chemicals. For example, amino acids are bonded with either carbohydrates, lipids, or nucleic acids depending on the body part of an organism. They can also be bonded with metal ions, hemoglobin, or chlorophyll. Spherical proteins such as enzymes and hemoglobin can be easily dissolved in water, having the ability to catalyze physiological reactions or transporting various materials. Proteins that are fiber shaped such as nails, toenails, or hair, have a rigid structure and do not easily get dissolved in water.

There are many different types of proteins that exist in a living organism. A structural protein (collagen), a storage protein (ferritin), a transport protein(hemoglobin), contractile proteins (actin and myosin), and membrane proteins are such types. Enzymes, hormones, antibodies, toxins are important proteins in the industrial field.

The structure of the protein is very complex, and the functions change depending on the structural characteristics. Peptide bond is made when condensation reaction occurs in between a carboxyl group of an amino acid and an amino group of another amino acid. This bond is tightly made so that it is not easily affected by pH or a solvent. However, a peptide bond can be broken apart by acids, bases, or protein-breaking enzymes. Protein structures can vary into primary structure, secondary structure, tertiary structure, and quaternary structure. Primary structure of a protein is a simple linear sequence of amino acids in a polypeptide chain. Secondary structure is long protein chains with specially-organized regions and can be either alpha helices or beta pleated sheets depending on where the region is. Alpha helices look intensely coiled, and beta pleated sheets are folded

Figure 4-1. Protein structure

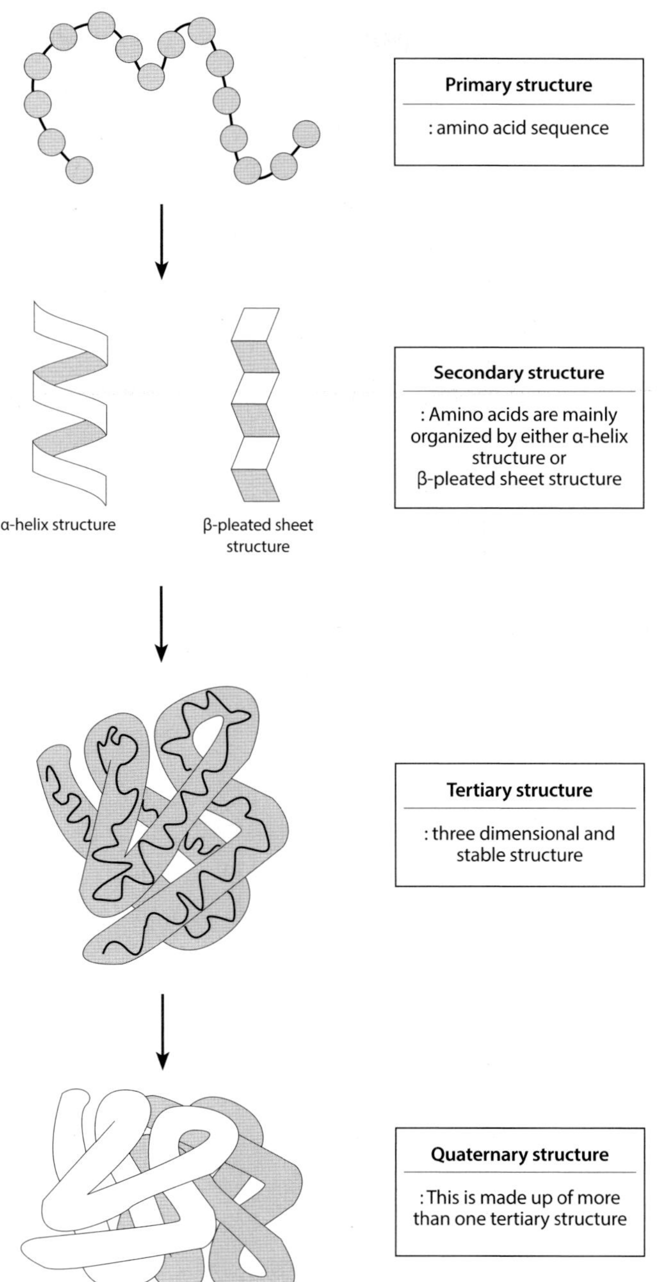

like papers. Many of sphere or fiber shaped proteins have this structure. Tertiary structure is generally found in spherical proteins that are three dimensional, and are fairly stable towards pH and temperature changes. The spherical proteins with tertiary structure are made up of hydrogen bonding, hydrophobic bonding, salt bridges, and disulfide bonds. Quaternary structure of a protein is made up of at least more than one polypeptide chain, and this plays a crucial role in building a protein structure from tertiary structures with many types of chemical bonds (Figure 4-1).

Now, let's look more into the basic concept, characteristics, types of the enzyme, and its use in the industrial field. The word enzyme originated from a Greek meaning of 'in yeast', and was named by a German physiologist Wilhelm Kühne (1837 – 1900) in 1878 to express the reaction of yeast during fermentation. Enzymes are about 5-20 nanometers in size, and are normally named by adding –ase at the end of the name. The names can also be expressed with the form of reaction. In here, substrate is a reactant or chemical substance that enzymes use. Enzymes are either composed of pure tertiary protein structures or with cofactors or coenzymes and proteins. Some of them are not functional if their cofactor and the coenzyme are not bonded together. Cofactor is an inorganic salt. Inorganic salts include Ca, Mg, Fe, Zn, Cu, K, Mn, Na, and act as ions. Coenzyme is

Figure 4-2. Enzyme structure

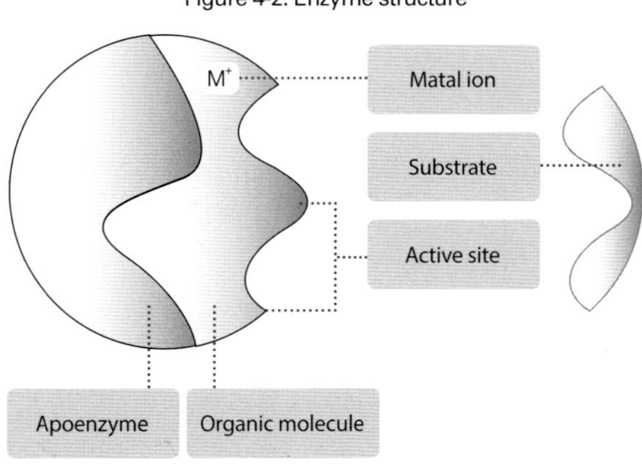

an organic material, and are usually complex vitamin derivatives such as NAD, FAD, Coenzyme A. As shown, enzymes have complicated multi-dimensional shape with many different types of bonds of various amino acids.

The protein part of an enzyme is called an apoenzyme. All apoenzymes have a part called the active site, and this active site is a critical site for reacting with substrates to become a different substance with new functions. The number of active sites vary by types of enzymes starting from one to many active sites. The main enzymes bonded with cofactors or coenzymes are called holoenzymes (Figure 4-2).

The enzyme only commits to one specific reaction due to the substrate specificity. It is composed of L- shaped amino acids, and substrates that are made specific for a certain active site that can only fit into that specific active site. This idea can be also described with Lock and Key theory which the lock can only be open by the key that is specifically made for that lock. In here, a lock is the enzyme active site, a key is the substrate, and the enzyme becomes active as a lock opens. This Lock and Key theory was first introduced by a German chemist Emil Fischer (1852 – 1919) in 1894. Fischer researched mainly on saccharides, enzymes, and proteins, and was awarded with the Nobel Prize for an achievement in the study of saccharide and purine synthesis. In 1958, an American biochemist Daniel Koshland (1920 – 2007) slightly modified the Lock and Key theory and suggested the Induced-Fit theory. This theory introduces the idea that the

Figure 4-3. Lock and Key theory of enzyme

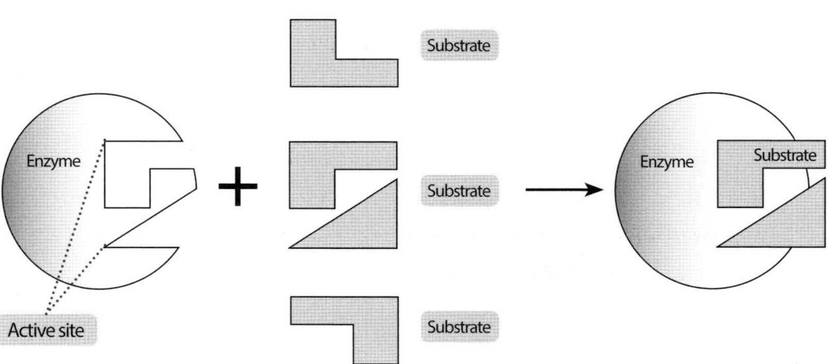

active site of an enzyme has a flexibility of being able to perfectly fit with the substrate for a precise and strong enzyme activation. These are the theories that well-represent the enzyme and substrate reaction. During the enzymatic reaction, an enzyme and a substrate react to each other and make an enzyme-substrate complex, and a substrate can be transformed into a product by detaching itself from the enzyme. The enzyme itself does not change its pattern, therefore, can react with the substrate again (Figure 4-3).

Normally, the performance of enzymes is tested by the maximum number of conversion of substrate molecule per unit time that a single catalytic site will execute for a given enzyme concentration under the optimum temperature and pH. This is called turnover number (TON) of enzymes and is counted by unit of kcat.

Let's look further into the types of enzymes. Enzymes can be classified into six different types. The enzyme classification of types and names are considered by the International Union of Biochemistry and Molecular Biology and are finally decided by the Enzyme Commission (EC). The followings are the types of enzymes.

1) Oxidoreductases

Oxidoreductases catalyze the oxidation and reduction reactions and cause hydrogen, oxygen atoms, or electron transfer into different substrate.

- Glucose oxidase: an enzyme that oxidizes glucose into gluconic acid.
- Alcohol dehydrogenase (ADH): an enzyme that breaks down ethanol (C_2H_5OH) into acetaldehyde (CH_3CHO).

Oxidoreductases are based on oxidation and reduction reactions, therefore, are mostly used in biosensors to detect specific materials.

2) Transferases

Transferases catalyze the transfer of a particular group from one molecule to another.

$AX + B = BX + A$

- Transaminase: an enzyme that catalyzes the transfer of amine group ($-NH_2$) of an amino acid.
- Transmethylase: an enzyme that catalyzes the transfer of methyl group($-CH_3$).

Transferases are mainly used in producing many types of oligosaccharides in food industries.

3) Hydrolases

Hydrolases catalyze the hydrolysis of a particular substrate.

$AX + H_2O = XOH + HA$

- Carbohydrase: an enzyme that hydrolyzes polysaccharides into mono or disaccharides.
- Protease: an enzyme that hydrolyzes protein into peptides and amino acids.

Hydrolases are widely used enzymes in the industries such as amylase, cellulase, protease, and lipase.

4) Lyases

Lyases catalyze the breaking of various chemical bonds (elimination reaction) and form a new double bond in atoms. These are different type of enzymes from hydrolases.

- Glutamate decarboxylase: an enzyme that breaks down glutamic acid into γ-aminobutyric acid (GABA) and carbon dioxide (CO_2).
- Pyruvate decarboxylase: an enzyme that breaks down pyruvic acid into acetaldehyde (CH_3CHO) and carbon dioxide (CO_2).

Lyases can break down the alginic acid in polysaccharides, which is the main ingredient of cell wall of brown algae.

5) Isomerases

Isomerases catalyze the conversion of a specific isomer to a different isomer, but only changes the isomer structure that is in the same molecular weight.

- Glucose isomerase: an enzyme that converts glucose into fructose.
- Arabinose isomerase: an enzyme that converts arabinose into ribulose.

Glucose can be converted into fructose by using immobilized enzymes in a packed-bed bioreactor.

6) Ligases

Ligases catalyze the combining of two large molecules with ATP by forming a new chemical bond.

$X + Y + ATP = XY + ADP + Pi$

- DNA ligase: an enzyme that facilitates the joining of DNA strand by catalyzing the formation of a phosphodiester bond.

New DNA strand can be made by DNA ligase from various organisms.

Next part is about the external factors that can affect the enzymes.

The main parts that can affect in enzymatic reactions are pH, temperature, and concentration of salt. These factors can affect immensely in enzyme activities, thus finding the optimum pH and temperature in enzymatic reaction is very important. Every enzyme has its own preferred temperature and pH for the best stability and maximum efficiency. The enzyme activity generally increases as the temperature increases, but the activity starts to drop as the temperature is increased over the optimum temperature range. The optimum pH of each enzyme works the same as temperature. Therefore, enzymes can't work properly if they are not under the optimum pH and temperature. The pH change especially directly affects in hydrogen bonds or salt bridges in enzyme structure. Some enzymes go through structure breakage or distortion that cannot be restored under strong acids or bases, and some enzymes go back to their normal function when the temperature and pH are changed back to the desired range. Using the buffer solution to change the pH helps with the stability of enzymes. Normally, enzyme activities can also be changed because of the types and concentration of the buffer solution. If the concentration of salts or organic solvents is increased too much, the enzyme reaction can be inhibited. Therefore, these concentrations should be maintained to the desired concentration.

Other than that, enzyme activities can be affected by the concentration of the substrate, enzyme or product, and the presence of an inhibitor. When the substrate concentration increases in the early stages, reaction rate increases. However, when it reaches its saturation, the reaction rate does not increase anymore. As a matter of fact, reactions may not even occur properly if the substrate concentration is too high due to the inhibition of enzyme reaction. This is called a substrate inhibition, which shows the importance of appropriate substrate concentration. If an inhibitor's structure is similar to the substrate's, the reaction cannot occur properly due to the competition between an inhibitor and the substrate to bind with the active site. Some inhibitors cause conformational change in an

Figure 4-4. Various factors affecting enzyme

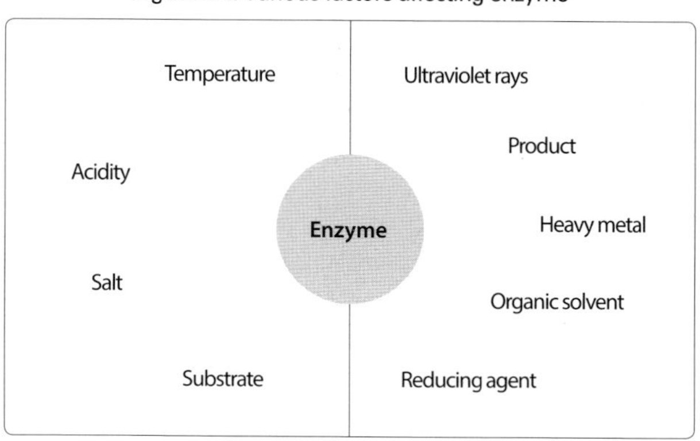

active site by binding to a different part of enzyme other than active site and inhibits the substrate to bind to that active site. These inhibitors are considered to be great candidates in developing new drugs. As an example, penicillin reacts with the enzyme for cell wall synthesis so that the cell wall cannot be synthesized anymore. A concentration of the product is also crucial. If the product concentration reaches the certain level in the reaction solution, it can cause inhibition of the reaction. This is called a product inhibition (Figure 4-4).

Now, let's look more into the reason why modern industries are switching their use of a chemical reaction over to an enzyme reaction. Many chemical reactions occur under high temperature and pressure with addition of organic solvents and catalysts, which give more complexity in processes. Also, use of chemical reaction brings in some negativities when considering the environment due to its large amount of byproducts produced with the main product. Enzyme reaction, on the other hand, uses fairly normal temperature and pressure with simpler processes, and makes less byproducts due to its characteristic in reacting to one specific substrate. Most of the enzyme reactions do not use organic solvents but rather use aqueous solvents, which is more economically friendly. Although the use of an enzymatic reaction brings in more positive advantages compared to a chemical reaction, its high price and stability are still questioned. If the enzymes are unstable, production expenses become too high compared to chemical reactions due to the expensive cost of enzymes. Therefore, active and proactive researches on increasing enzyme stability and its immobilization technology for multiple uses are crucial in the industrial field (Figure 4-5).

Figure 4.5. Methods for enzyme immobilization

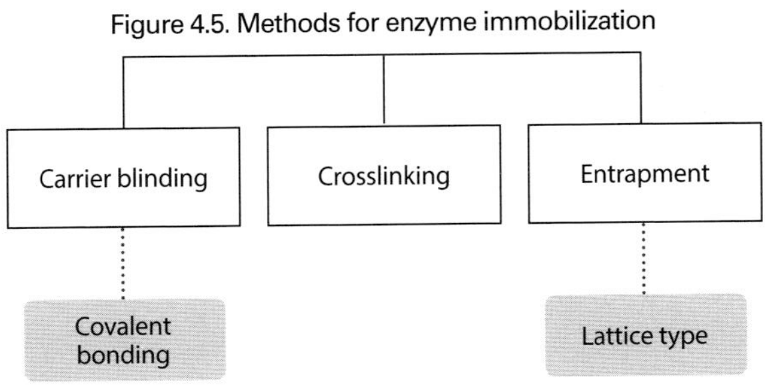

Figure 4-6. Structures of penicillin and 6-APA

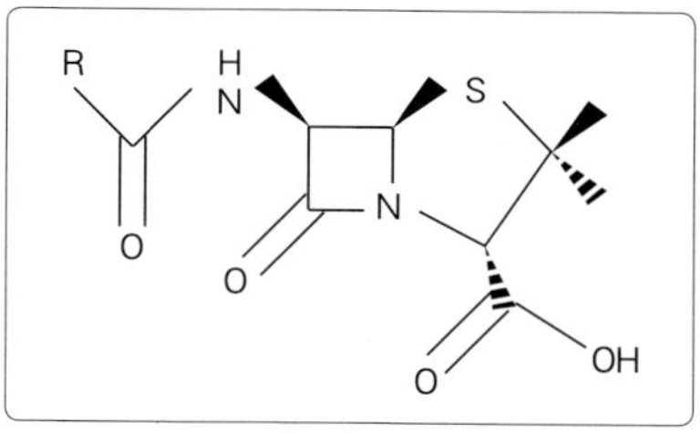

Penicillin

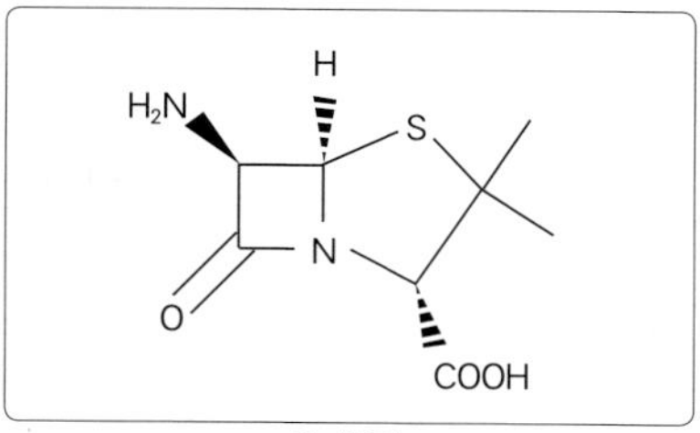

6-APA

As the environmental safety has become more important, many processes for chemical synthesis are substituted to the enzymatic processes. In here, a 'process' is a series of procedures of a substance becoming a product. A good example would be 6-aminopenicillic acid (6-APA) that is made from penicillin. As a matter of fact, penicillin itself cannot be used effectively now because most of the pathogenic bacteria has developed resistance against it. Though the pure penicillin cannot be used now, different types of penicillin have been chemically modified to fight against many pathogenic bacteria. 6-APA is an important intermediate substance to change the chemical structure of penicillin. 6-APA used to be synthesized through complex chemical processes in the past, but is now synthesized through the bioconversion process of penicillin by immobilizing an enzyme called penicillin acylase. This is how penicillin should be constantly produced as a basic substance through microbial cultivation (Figure 4-6).

Although we previously mentioned about the reaction between enzymes and substrates, let's assume that enzymes and substrates are both dissolved in the enzyme bioreactor. The bioreactor is maintained to have the proper temperature, pH, and concentration of enzymes and substrates for efficient enzyme reaction, then mixing the solution of enzymes and substrates together in a constant agitation speed. Some of them will react together and become a final product, and unreacted ones will constantly react together to make product. After most of substrates were converted to the products, this batch reaction would be finished and followed by separation of final products. In this case, it is quite difficult to reuse the enzymes because they were mixed with the products and the unreacted substrates. It is quite challenging to separate the enzymes to be used again because of expensive separation cost. Together with that, enzyme production cost should also be considered. Therefore, recycling enzymes that are still functional is crucial in saving energy, time, and money. This is exactly why immobilization technology was developed (Figure 4-7).

Figure 4-7. Enzyme bioreactor for continuous bioconversion

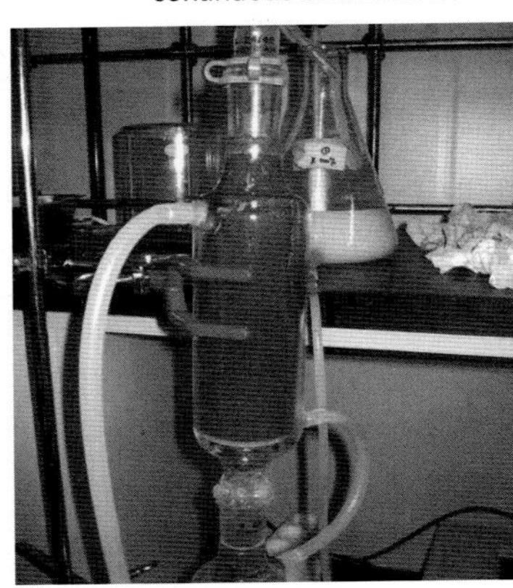

Enzyme immobilization technology was developed either by adsorbing or by binding an enzyme to an inert and insoluble material (carrier). As immobilized enzymes are attached to the inert and insoluble materials, they can be easily segregated from unreacted substrates and products within the bioreactor and be reused. Although the reusable cycles can vary depending on the stability of immobilized enzymes during industrial enzymatic processes, they could be used hundreds of times. As mentioned before, immobilized enzymes can be classified into industrial enzymes and therapeutic enzymes. Therapeutic enzymes are classified into diagnostic enzymes and medical enzymes. Many industrial enzymes are used as immobilized enzymes in medicine or food field which develop 6-APA-related medicine in pharmaceutical industries or converting glucose into fructose in food industries. An enzyme that plays in the reaction of glucose to fructose conversion is glucose isomerase. Immobilization technique is also used in the application of diagnostic enzymes. As an example, oxidases or reductases are normally immobilized and attached on the blood sugar measuring kit. Though many types of kits are on the market, most of them share the same idea of immobilizing these enzymes

Figure 4-8. Industrial application of enzymes

Industrial enzymes	Diagnostic enzymes	Therapeutic enzymes
Food, Detergent, Textile, Chemical industries	Biosensors, In vitro diagnostic (Most of oxidation/reduction enzymes)	Digestant, Thrombolytic agent, Coagulase, Antitumor enzyme

on a thin film. Medical enzymes are normally applied as injections or oral intakes, but the drug delivery system with the basis on immobilization technology is currently being emphasized in oral intakes due to the importance of safe delivery of drugs to the desired treating area in the body without chemical destruction (Figure 4-8).

Chapter 5

Fascinating Procedures for Various Biotechnological Products

Chapter 5 explains the general idea on biological processes for the application of organisms and the enzymes produced by them. The concept of biotechnological studies stand on the process of various biotechnological products produced by using microorganisms, animal and plant cells, or enzymes through collaboration between universities, research institutes, and industries. Although active research is involved in this field, a successful product can be produced only when the upstream technology with its foundation on biology, microbiology, biochemistry, molecular biology, and molecular genetics and the downstream technology which is based on chemical engineering are well-balanced between each other. Upstream technology is mainly used to find out functions and mechanisms of an organism and to develop industrial strains with this information. Downstream technology focuses on bioprocess, which includes mass production and bioseparation of products. Some researchers organize with more details as upstream technology, midstream technology, and downstream technology. In here, midstream technology is the technology of mass production of a product from the raw material. Downstream technology includes bioseparation process technology. All technologies mentioned above are important to flow all together as the river water flows from upstream to downstream. A problem from a single stage may occur a serious difficulty in producing products. Ultimately, each part of bioprocess is actively under research, and are divided into strain development, mass cultivation of strains, and separation of a product.

Figure 5-1 shows the systematic flow chart of the bioprocess to produce

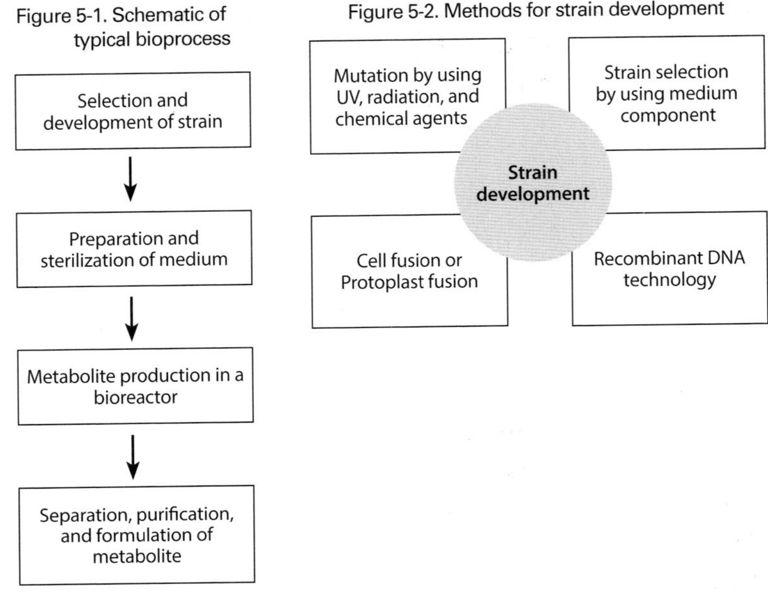

biotechnological products. Biochemical engineering field is especially crucial in developing the bioprocess itself and mass production of useful products by scale-up. Not only that, as high value-added products such as recombinant proteins have developed, many researchers have been working on developing various types of bioprocesses. Therefore, the suitable systems should be developed and exploited by studying each bioprocess qualitatively and quantitatively.

Strain development, an upstream technology, has a main purpose on increasing a cell activity to improve yield and productivity of a product for better economic feasibility in industries. The wild type strain found from soil or a specific environment is normally low in yield and productivity due to its low activity. The followings are some ways to increase a cell activity (Figure 5-2).

- Strain selection according to medium composition
- Mutation by UV, radiations and chemical agents
- Cell fusion
- Recombinant DNA technology

If these methods are used, a strain with an excellent activity could be developed.

Much effort in manpower and constant strain selection is needed in order to develop an excellent industrial strain from a wild type. The simplest that the researchers can do in a laboratory is to constantly change the composition and concentration of a medium to observe which strain fits the best as an industrial strain with its desired growth rate, pigmentation and concentration of a product. For example, Korean rice alcohol drink called Makgeoli, is made of rice by using mold (filamentous fungus). This fungus produces amylase to break down rice into glucose and uses it as a source of energy. If this fungus is introduced onto a petri dish with the minimum amount of nutrient to survive, some of the stronger colonies can be observed with their increased amylase activity. In a different word, a fungus that only lived in an environment with abundant nutrient may not survive under an extreme environment. Therefore, the possibility of finding an excellent colony increases if the colony can survive under the extreme environment with less nutrients.

Although using mutation of microorganisms and plants by UV rays, radiation (x-ray, γ-ray, proton beam, ion beam), and chemical components is the traditional way, this method is still popular within the industries to produce biotechnological products. As mentioned in Chapter 2, penicillin, the most well-known antibiotics, was luckily found by Alexander Fleming with serendipity. *Staphylococci* growth on the medium added with agar and nutrients was found to be inhibited by the material released from the contaminated fungus. This compound was penicillin produced by the fungus *Penicillium notatum*, which is a green mold. Though Fleming discovered penicillin, the fact that other researchers made a clue for commercialization of penicillin right away for better future of mankind can be highly valued. The first fungal strain used in the past had very low productivity of penicillin. However, the productivity has immensely increased after the selection of better strains with high activity and using the mutation method. Another important fact that the researchers found along with the high activity was that the nutrient includes a special material with a penicillin intermediate that can also increase the productivity of penicillin. This shows that nutrient composition is also crucial for higher productivity of penicillin. Other than these two factors above, having the perfect culture conditions and developing new culture methods also helped to increase penicillin productivity level.

Using microorganism or plant/animal cells give great potential in creating exceptional cells by using cell fusion and recombinant DNA technology. A simple example of using cell fusion would be developing yeast cells for good beer fermentation. This can be also called protoplast fusion. Let's say the yeast Group A has a great flavor but lacks in ethanol production, and Group B's ethanol production is great but the flavor is missing. This is when the industrial laboratories use protoplast fusion to create the perfect new group of yeasts that fulfill both the flavor and the ethanol level. In detail, yeast A and yeast B's cell walls are eliminated by an enzyme. Then, these two cells can be fused together by using polyethyleneglycol (PEG) so that the new cell wall would be built later. After the new fused cells grow, the researchers only look for yeast colony with good flavor and alcohol level. Though the protocol seems easy, development of protoplast fusion takes large amount of time and effort. Plant and animal cells are being developed with the same method. Plant cells are fused together to make a hybrid plant cell. For animal cells, a normal animal cell and myeloma cell are fused together to develop a such high value-added product like monoclonal antibody. In 1975, a German biologist Georges Köhler (1946 – 1995) and an Argentine biochemist César Milstein (1927 – 2002) developed a new technology of producing monoclonal antibody from a hybridoma cell that was made from a B cell and a malignant tumor cell through PEG fusion. With this innovative cell fusion technology, they obtained the Nobel Prize in Physiology or Medicine in 1984. Unlike polyclonal antibody, monoclonal antibody contains only one type of antibody.

Recombinant DNA technology is mainly used in application of developing useful materials by injecting genes of a microbial cell into another microbial cell DNA that is under different genus or species or adding DNA sequences of an animal/plant cell into another animal/plant cell DNA sequences. In 1973, an American biochemist Stanley Cohen (1922 –) and the co-founder of Genentec, Inc., Herbert Boyer (1936 –), developed the recombinant DNA technology which DNA can be cut and pasted back together by using restriction enzyme and ligase. With this technology, they successfully recombined new DNA sequences by inserting the foreign DNA into a bacterial DNA. Using this technology on the microbial DNA is possible because they contain extra circular DNA source called plasmid. Stanley Cohen obtained the Nobel Prize in Physiology or Medicine in 1986 with an Italian neurobiologist Rita Levi-Montalcini (1909 – 2012) for developing a nerve growth factor. Herbert Boyer developed synthetic insulin by using a recombinant bacterium in 1978, and produced growth hormone in 1979.

The most popular microbial cells used are *E. coli*, *B. subtilis*, and yeast. The most ideal idea would be economically producing only desired products by inserting the desired genes from different microbial species or animal or plant cells into a plasmid of those cells. Although this is ideal, sometimes the expression of the introduced gene is not performed properly. In this case, product should be produced with different methods or using animal or plant cells. Restriction enzyme functions as scissors to cut certain parts of genes in a plasmid, and a ligase pastes the plasmid gene and the inserted foreign gene together. A plasmid inserted with a foreign gene is called a vector or a vehicle. This vector is transformed into a suitable microbial cell and only the cells with foreign gene properly inserted are chosen. As mentioned in Chapter 2 briefly about the case of recombinant insulin, human insulin gene was inserted into *E.coli* gene and was successfully mass-produced and commercialized with the approval of the FDA in 1982.

Figure 5-3.
Methods for mass production of product

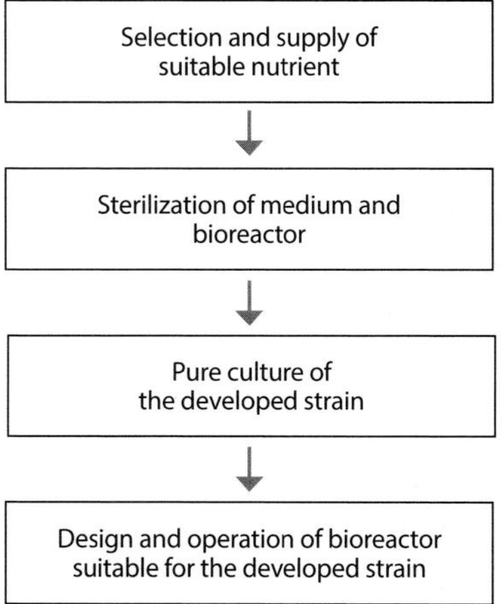

A midstream technology, mass culture of a cell, is a vital step for mass production of useful material. Mass production of useful material is possible only when cells are in high amount or concentration. This mass culture can be performed in various bioreactors with the cell's biological reactions depending on its characteristics. Desired biological reaction can take place if the environment of a bioreactor is kept constant such as temperature, pH, or nutritious factors. As mentioned in Chapter 3, a cell itself can be considered as a small, independent factory. If the factors that play a role in useful material production are carefully controlled, more than a hundred million 'factories' can be cultured inside the bioreactor, and mass production of the useful material will be successfully operated. The following list is important factors that need to be considered during midstream technology (Figure 5-3).

- Selection, preparation, and pretreatment of raw materials containing carbon or nitrogen source in a medium.
- Complete sterilization of all media and bioreactors to prevent contamination.
- Proper design and operation of the bioreactor.

Nutrition is crucial for thriving growth of cells. Jacques Monod (1910-1976), a French biologist, explained clearly about the correlation between cell growth and nutrition, and a growth rate in cell culture. Monod was prized with the Nobel Prize in Physiology or Medicine for finding genetic control of enzyme and virus synthesis. He also wrote his doctorate thesis about bacterial growth in a culture medium containing two types of saccharides. He is also famous for his Monod equation, which is a mathematical model for microbial growth.

Laboratory cultures generally use more expensive pure nutrients supplied from the culture medium manufacturing companies for better understanding of a cell's characteristic. However, price should be more thoroughly considered and industrial culture medium should be used ultimately. For industrial medium, choice of carbon and nitrogen sources are crucial in minimizing problems such as an oxygen transfer, viscosity of culture medium, foam, or pH changes during culture. Although product yield and productivity are crucial when using industrial medium, minimizing by-products that cause problems in downstream process is also very important. When the chemical structure of the by-product is similar to the desired product, it is especially challenging to separate the by-product. Due to such high production cost, gratification in economical perspective may fail. Not only that, all processes may be completely stopped if supply fails even though industrial medium fulfills all desired parts. In recent years, many bioindustries cultivate and pretreat biomass at low cost in other countries, and transport into the homeland. All factors mentioned above should be thoroughly considered to pick the right raw material.

Contamination prevention is very important in mass culture. If cultures or expensive recombinant proteins are contaminated by other microorganisms during mass production, quite a lot of time, effort, and cost become wasted. The basic idea of sterilization was introduced by a French chemist and a microbiologist, Louis Pasteur (1822-1895), after he developed pasteurization. Louis Pasteur is known for his work in vaccines, microbial fermentation, and low-temperature sterilization (pasteurization). He had managed the whole Pasteur Institute since 1887 until he was deceased.

If milk is heated up to 65℃ for 30 minutes, all zoonotic pathogens can be annihilated. By using the inverse relationship between temperature and time, time can be much saved when temperature is increased. This idea is especially used in laboratories during cell culture. Though volume difference may bring outliers, most of culture media in the Erlenmeyer flasks can be sterilized under 121℃ for 15 minutes. Solid cultures have higher possibility of contamination than liquid cultures, so sterilization conditions should be adjusted so that heat transfer is smooth. On pilot-plant or industrial scale, sterilization is performed with hot steam under high pressure because the volume of the bioreactor becomes very large.

A reactor is a container capable of performing a chemical reaction. Therefore, a reactor that is capable of performing a reaction by a living cell or an enzyme secreted by a cell is called a bioreactor or an enzyme bioreactor. Since the ancient times, it is presumed that various containers have been used as mankind made cheese, yogurt, liquor and such. These containers can be regarded as various bioreactors. There are also naturally occurring bioreactors that can be found around us. When edible fruits are accumulated in between the grooves of a thick tree in the forest and the proper temperature and humidity are maintained by the rain stuck in the groove, fermentation occurs by the yeast attached to the fruit skin, and ethanol is produced. The groove of this tree is considered a natural bioreactor.

The actual use of bioreactors is closely linked to the wars such as World War I and II. During World War I, production of acetone was crucial due to its important chemical composition in making explosives in the British military industry. The British chemist Chaim Weizmann (1874-1952) successfully developed acetone-butanol-ethanol, 3:6:1 (ABE) fermentation by *Clostridium acetobutylicum* using corn as a medium, which led to a successful mass production of acetone. This is one of the first application of an industrially developed bioreactor. Weizmann was a Jewish British, elected to be the first president of Israel in 1948, and is well-known as the president who produced acetone.

One of the representative use of a bioreactor in the earlier times of bioindustries was saving many lives by mass production of penicillin during World War II. Jasper Kane (1903-2004), an American biochemist who was working at Pfizer, performed deep-tank fermentation of penicillin with a bioreactor which successfully produced penicillin in industrial scale. Figuring out how to distribute massive amount of medicine through mass production is as important as developing new medicine for diseases.

As mentioned above, a bioreactor is designed to sterilize with steam for contamination prevention. Each bioreactor is designed differently depending on the characteristic of a cell or process system. Mass production of all biotechnological products is challenging without constant development of a bioreactor, so various types of bioreactors are being designed until now. The next chapter will take more in depth about basic principle and types of bioreactor.

A downstream technology, separation of products, is a lot more expensive than the upstream or midstream technology due to more complex and more required steps for product separation and purification. It is challenging to separate only the desired substance because many different types of components are produced by the cell and they are generally present at a low concentration in the aqueous solution in a bioreactor. Therefore, various separation methods are used. First and foremost, cheaper method option should be selected. Minimizing the number of bioseparation steps is also important since bioseparation processes are so complex that the product can be easily damaged. Filtration, centrifugation, and extraction are frequently used in the initial stage of separation, but various chromatographic methods are mainly used in the latter stages for complete purification. The following list is the popular methods used in bioseparation, and the description of each will be explained in Chapter 8 (Figure 5-4).

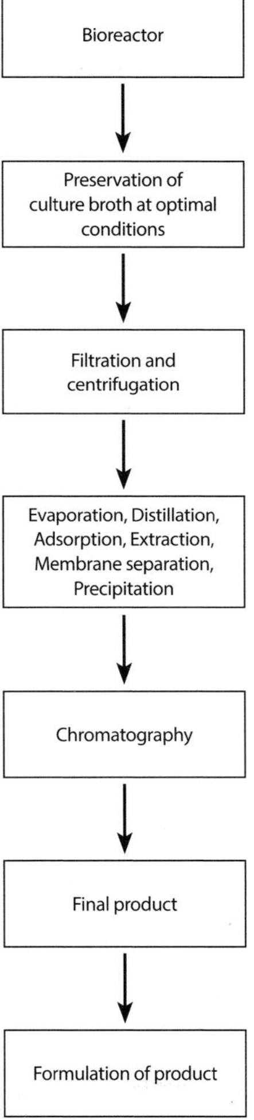

Figure 5-4. Methods for a typical bioseparation

- Filtration, centrifugation
- Adsorption, extraction
- Membrane separation, evaporation, distillation, precipitation
- Chromatography

Formulation is the process of making a separated product into a final product when the desired product is produced from a bioreactor and is separated by an appropriate method. The product must be stabilized for a long time through formulation, and the formulated product is sold and dispersed throughout domestic and overseas markets in a liquid or solid state. So far, upstream technology, midstream technology, and downstream technology were briefly described.

When appropriate technologies were developed from upstream to downstream technologies to economically produce the final product, midstream and downstream technologies must be scaled-up for mass production. Scale-up is an essential process for the mass production of biotechnological products. Just because the optimization of culture conditions and bioseparation methods in laboratory scale was successful does not mean they can be directly applied in an industrial scale. A technology can be operated in an industrial scale only when all conditions for production and separation are successful under the intermediate or pilot scale, which is the mid-scale between the lab scale and the industrial scale. If problems arise in culture conditions or separation

Figure 5-5. Scale-up process from laboratory scale to industrial scale

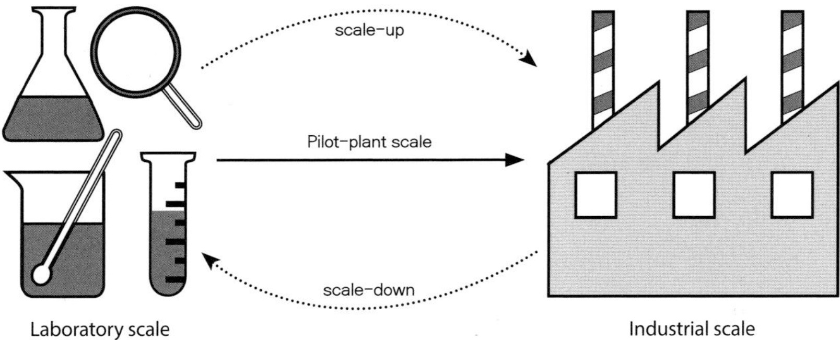

methods, problems must be identified and resolved by returning from the intermediate/pilot scale to the lab scale or from the industrial scale to the intermediate/pilot scale. This process is called scale-down (Figure 5-5).

Here is an example for a better understanding on the overall flow of biological processes. When you go to the market, you can easily buy the Baker's yeast and the Brewer's yeast. Before going over the processes of producing the Baker's yeast, we will go over the history and characteristics of yeast briefly.

The word bread originates from a Portugese word pão, which changed pronunciation into 'pang' in Japanese into Korean. Fermentation of bread using yeast is thought to be started since the ancient Egypt period. Ever since then, yeast has been closely related with human life. Fermenting yeasts are normally within the size of 5-10 ㎛and are either spherical or ellipsoid shaped. The scientific name of a general fermenting yeast is *Saccharomyces cerevisiae* (Baker's yeast). In the early times, *S. cerevisiae* that was used for fermentation was wild type. Including yeast, other types of fungi exist such as mold that grows with thread-like patterns and mushrooms that form fruiting bodies. Yeast propagation occurs through two different types: sexual reproduction and asexual reproduction. Asexual reproduction includes budding and fission of yeast. Budding is the most well-known form of yeast reproduction. A daughter cell starts to grow on the surface of the mother cell, and separates itself from the mother cell as the daughter cell grows big enough. A scar can be observed under the microscope where a daughter cell is separated on the surface of the mother

Figure 5-6. Production of Baker's yeast

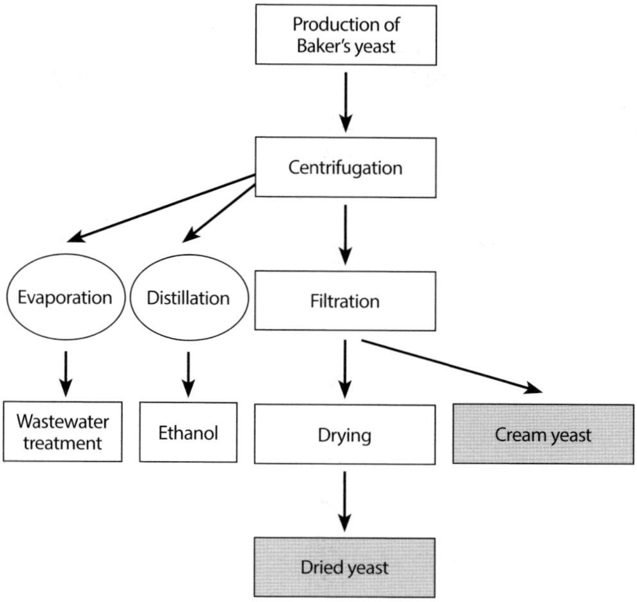

cell. Some other types of yeasts are fission yeasts which yeast cells divide by medial fission. Sexual reproduction of yeast occurs by two haploid cells fusing together to make a diploid cell with two sets of chromosomes. A diploid cell's nucleus divides and forms an ascospore, the ascospore develops and germinates, and become two haploid yeast cells. Sexual reproduction of yeast could be used for improvement of fermentation ability or storage.

Based on the basic principles of bioprocess mentioned in previous paragraphs, upstream, midstream and downstream technologies are needed to produce Baker's yeast (Figure 5-6). Upstream technology is strain development which is the process of selecting yeasts that have only strengths by cultivating yeasts with different characteristics or artificially fused cells to improve bread fermentation efficiency or storage stability of yeast. Once an excellent yeast strain has been selected, it can be purely cultured to obtain an appropriate amount of yeast, and then thousands or tens of thousands of seed cells can be prepared, freeze-dried, and stored in a deep freezer or they can be mixed with glycerol and kept frozen for a certain period of time. The yeast strains must be preserved in these ways so that they can be used at any time, especially when contamination occurs during culture on petri plate and bioreactor in laboratories.

In the midstream technology, cheaper raw materials must be selected first. One of the cheapest raw material known is molasses which is a carbon source. Molasses can be obtained by concentrating the remaining material after extracting sucrose from sugarcane or sugarbeet and has dark brown color containing 50-60% of sucrose with turbidity and high viscosity. The concentrate can be diluted and used as a carbon source for yeast culture. If cheaper but highly concentrated nitrogen source is needed, corn steep liquor (CSL) may be the best choice. Corn steep liquor can be obtained by extracting starch from corn and concentrating the remaining material. It contains a high concentration of nitrogen source including various amino acids and can be appropriately diluted and used as a nitrogen source. These materials can be sterilized and unneeded materials can be removed with a centrifuge if the impurity concentration is high.

After choosing a suitable bioreactor for culturing yeast, sterilization of these nutrients is done. After that, sterilized nutrients go into the bioreactor, adjusting with the optimum temperature and pH, and then seed cells are being inoculated. A proper amount of yeast cells will be obtained within the certain time period, and the yeast itself is considered as a product.

Downstream technology generally uses filtration or centrifugation to collect yeast produced in a bioreactor. Yeast is first separated by a centrifuge, stored in a storage tank that is maintained at a constant temperature, and the yeast cake is collected by using a rotary drum filter. The collected yeast is made into two types of products. One is packaged raw as wet yeast and sold to the domestic markets. The other is dried with a dryer as the dry yeast and is sold to the overseas markets because of its long expiration date.

The main product is Baker's yeast, but ethanol is also produced as a by-product during the cultivation in the bioreactor. Ethanol is contained in the remaining liquid after all yeasts are collected by filtration and centrifugation. Therefore, anhydrous ethanol can be produced and sold through distillation and concentration. The remaining liquid go through wastewater treatment. As upstream, midstream, and downstream technologies are continuously performed, yeast can be manufactured as a product and be

Figure 5-7. Various biotechnological products

Cell	Extracellular product	Intracellular product
Baker's yeast, Brewer's yeast etc.	Antibiotics, Amino acids, Organic acids, Enzymes, Alcohols etc.	Recombinant proteins etc.

sold in the markets. In here, only rather simpler examples are shown that are easier to manufacture, such as cells themselves being products. The process of collecting product is a lot more complicated if the product is contained in liquid or accumulated in the cell. Also, the degree of purification depends on the purpose of the product use. Products that are administered to humans, such as biopharmaceuticals, are paid attention more carefully because they can cause toxic reactions in the presence of impurities in the product.

To give an example of an enzyme, industrial enzymes are mostly partially purified with different purities from industry to industry. Diagnostic enzymes need to be purified up to 95-98% in order to be used. Therapeutic enzymes should be purified close to 100% to be used as therapeutic medicine.

All biotechnological products are made through a complex process that is optimized based on upstream, midstream, and downstream technologies. Well-known products include organic acids such as citric acid, amino acids, antibiotics such as penicillin, industrial enzymes such as amylase, protease, cellulase. Diagnostic enzymes such as glucose oxidase, and therapeutic enzymes such as antibodies, insulin, thrombolytic agents, blood coagulants are popular biopharmaceutical products (Figure 5-7).

Generics and Biosimilar, which are mostly cited in newspapers, are also being produced based on these bioprocesses. Generics is a so-called copy drug, originally developed as a new drug from one company and produced for about 20 years, and then the other company develops the production technology before the patent expiration and produces the same chemical structure after the expiration of the patent period. Due to the exactly same chemical structure, they do not need to be approved by the FDA, but they must be approved by the country for their entire manufacturing process. On the other hand, biosimilars are mainly protein drugs, and it is very difficult

Figure 5-8. Biopharmaceuticals which patents have been expired in 2015~2017.

Product	Ingredient	Efficacy	Patent expiration
Neulasta	Pegfilgrastim	Promotion of leukocyte	2015
Lantus	Insulin glargine	Diabetes treatment	2016
Tamiflu capsule	Oseltamivir phosphate	Treatment and prevention of influenza virus infection and avian influenza infection	2017
Vytorin tab.	Simvastatin, Eztimibe	Hyperlipidemia treatment	2016
Humira	Adalimumab	Rheumatoid arthritis treatment	2016
Paxil CR tab.	Paroxetine hydrochloride	Depression treatment	2016
Iressa tab.	Gefitinib	Lung cancer treatment	2016
Tygacil inj.	Tigecycline	Antibiotic	2016
Baraclude tab.	Entecavir	Hepatitis B treatment	2015
Avodart tab.	Duasteride	Treatment of hair loss, and prostate enlargement	2016

to make the exactly same chemical structure even if the manufacturing technology is developed and prepared properly. Although the structure and function are similar, toxicity may arise even if few amino acid sequences are different. Therefore, they must be approved by FDA as if getting approved for a new drug. In terms of efficacy, more improved drugs than the original biosimilar may be developed, and they are called biobetter. These biosimilars and biobetters are called biophamaceuticals because they are artificially produced substances that are originally produced in animal or human body. Biophamaceuticals have higher possibility of success compared to conventional synthetic drugs and takes less time and cost (Figure 5-8).

The new drug is defined as a completely newly made drug that has not been released to date. Therefore, it can be recognized as a new drug only when it has originality in the chemical mechanism, has a new chemical structure and covers the problems of the original product, or has to be remarkably improved in terms of its efficacy and stability. These new drugs would be called the innovation drug or the first-in-class drug. Developing new drugs takes time and money even for the countries with developed medical biotechnology field, so many companies focus mostly into me too drugs or follow-on drugs. Follow-on drugs have similar mechanism as innovative drugs, but with partially modified chemical structures or improved formulation for higher functionality and efficacy. Normally, the drugs that can only be obtained by a doctor's prescription are ethical-the-counter drugs (ETC) and the ones available at drugstores are over-the-counter drugs (OTC).

The process of new drug development is quite complex. First, basic research based on medical background about a certain disease is needed for possible ways to overcome the disease. Then multiple theories are set from innovative ideas and technologies, and new materials through chemical synthesis, microbial fermentation, and extraction from animals and plants should be obtained. When the new candidate materials are found, they are tested for preclinical trials such as toxicity, pharmacology, and pharmacokinetics through Good Laboratory Practice (GLP) before they go into clinical trials. Once the new materials are proved with safety and effectiveness, they are used for clinical trials on human body under the control of GLP. Clinical trials have phase I through III studies, and each phase must submit clinical protocol and test report and be approved by Good Clinical Practice

Figure 5-9.
The drug development process

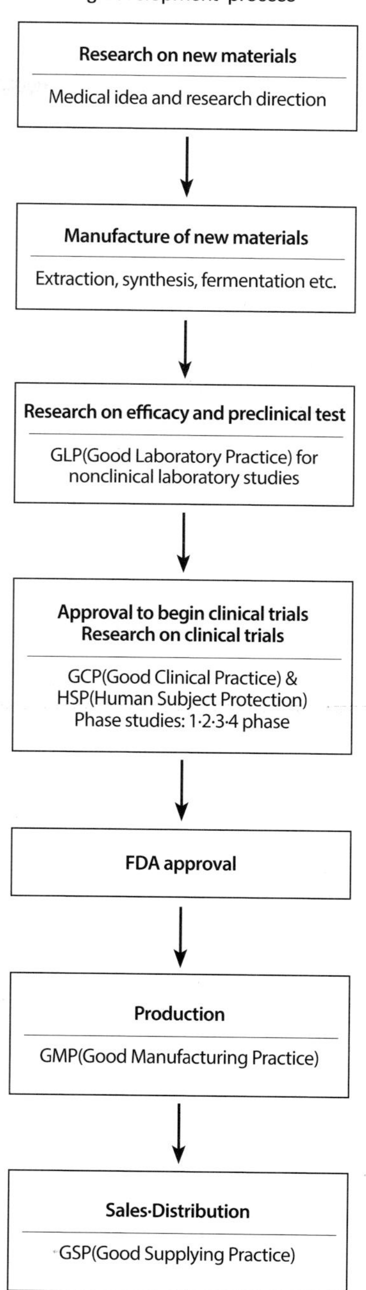

(GCP). Phase I studies focus on safety of the drug with its pharmacokinetics and pharmacology, and the drug is usually applied to 20-80 healthy human subjects. Phase II studies focus more on the effect, safe dosage and usage of the drug, and hundreds of human patients are injected with it. By Phase III clinical trials, the effect and safety of the drug is mostly proven through Phase I and Phase II. After certainty is established, hundreds to thousands of patients are injected with the new drug.

As the effectiveness and safety are proven through these processes and are finally authorized by Good Manufacturing Practice (GMP), the new product starts to be produced. In here, international legal standard is Current Good Manufacturing Practice (CGMP) and Korean standard is Korea Good Manufacturing Practice (KGMP). As the new product is made after the authorization, the product is circulated throughout the markets under permission of Good Supplying Practice (GSP). After circulated on the market, Phase IV clinical trial is conducted to double-check the safety of a product (Figure 5-9).

New drugs go through all the procedures above, and become distributed world-wide after being mass-produced from a well-organized bioprocess.

Chapter 6

Bioreactors for Artificially Cultured Organisms

The name bioreactor is a compound word originated from a reactor where biomaterials such as cells or enzymes produced by cells are reacted in a reactor. The container that runs these whole processes of biological reactions is called a bioreactor. Originally, reactors were mainly used in the areas of chemistry and chemical industries for converting various raw materials into desired chemical products. In the early stage, it has been used for basic experiments. As the necessity of mass production increased, reactors were upgraded into larger scales as an indispensable tool for various industries.

As the field of biotechnology has developed, a variety of bioreactor types were needed to produce various biotechnological products. Biomaterials include microorganisms, plant cells, animal cells, and enzymes. In general, a container that is needed for production of cells or metabolites by mass culture of cells is called a bioreactor, and a container that produces biotechnological products by carrying out bioconversion reaction using enzymes or immobilized enzymes is called an enzyme bioreactor.

There are many different types of bioreactors used in the biotechnology fields (Figure 6-1). Bioreactors range from laboratory scale containers such as test tubes, conical flasks, and smaller scale bioreactors to large scale bioreactors used in the biotechnology industries. In fact, there are many types of bioreactors used on general culture such as a stirred-tank, a fluidized-bed, a packed-bed, and a membrane bioreactor. These types of bioreactors can be selected or newly designed upon the suitable use in different applications in biomaterials or systems. In addition, different types of cell culture methods in a bioreactor can be chosen depending on

Figure 6-1. Various bioreactors

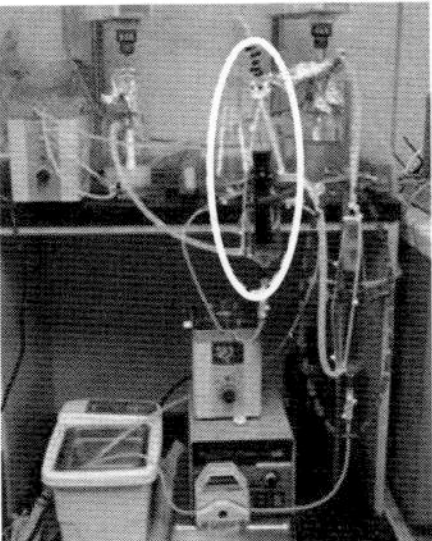

Packed-bed bioreactor

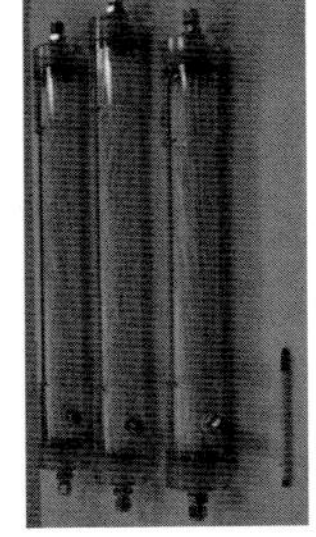

Membrane bioreactor

Fluidized-bed bioreactors

Stirred-tank bioreactor

the operating method. A batch culture ends its reaction process after one cycle. Continuous culture can produce a bioproduct by using a biological system continuously. Fed-batch culture feeds the substrate with intermittent or constant rate. Most of the industries use batch or fed-batch culture methods for mass production of biotechnological products. A continuous culture method is not as popular due to its higher possibility of contamination in the connected lines on the bioreactor. Though, this method is used in ethanol fermentation or wastewater treatment where contamination is not a problem. Therefore, continuous culture has high potentials of being used more with higher productivity if the technical problem in regards to contamination is resolved. In particular, the design of the bioreactor changes depending on the characteristics of the cell culture. Cell culture can be divided into the liquid culture or the solid culture depending on the state of nutrients. Liquid cultures perform in an environment containing liquid nutrients, and solid cultures perform under environment with solid

Figure 6-2. Liquid culture(Left) and Semi-solid culture(Right)

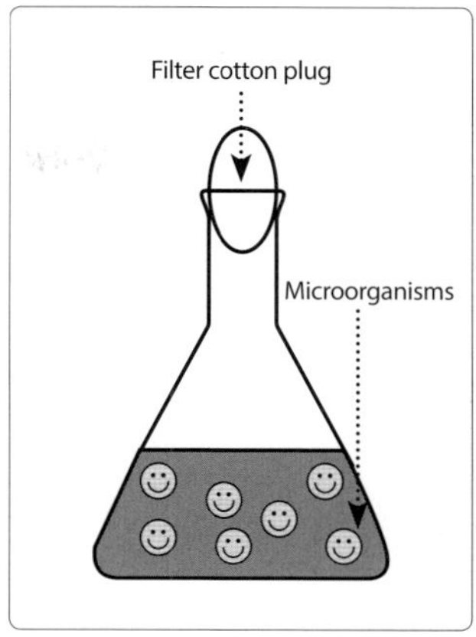

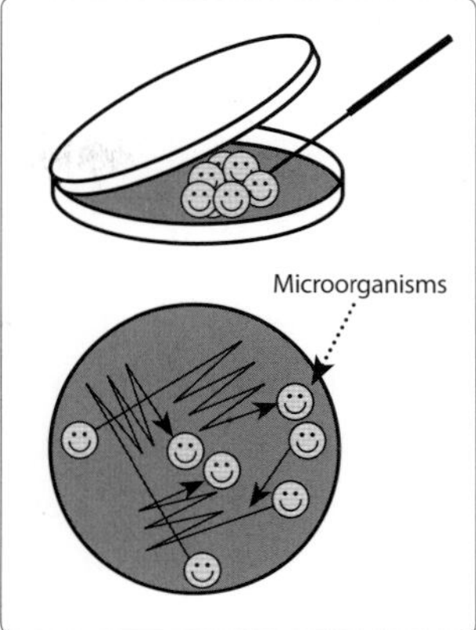

nutrients. Also, depending on the oxygen needs, containers are constantly shaken or static at one place during culturing (Figure 6-2).

An example of food fermentation is manufacturing rice wine (rice Makgeolli). Rice wine is first prepared by cooked rice mixed with fungus that produces amylase in a fixed volume of container under a certain temperature. As fungus starts growing on the surface of rice, amylase is produced, and amylase breaks down rice into saccharides. In here, cooked rice is a solid nutrient, and rice and fungus are mixed well together for the fungus to have an access to oxygen for amylase production. Therefore, this whole process is considered as an aerobic solid culture. In the next step, the cultured yeast is added and mixed to the degraded rice in a different container. As yeast is added to the mixture, fermentation starts to occur. In this process, aerobic liquid culture method or submerged culture method is used in a separate container to increase the concentration of yeast. Then, the rice wine fermentation occurs at the final stage in a big container. Since this process does not require oxygen supply, it is a static liquid culture. All the containers used in these processes are all bioreactors.

In laboratories, a bacterial growth can be often observed on the surface of petri dish made of solidified mixture of nutrients and agar after the strain is applied on the surface of petri dish and incubated inside the culture incubator. This type of culture is called solid or semi-solid culture. If you go to the countryside of Korea, you may have seen 'Meju', a solidified fermented soy bean paste, hanging under the roof with a rope made of rice straw. Meju is often kept with the rectangular or spherical shape to effectively perform solid culture. The rope contains many types of microorganisms including *Bacillus subtilis*, therefore, plays a role as seed bacteria. This is also another good example of traditional solid culture.

Another good example of traditional food fermentation is 'Sik-hye', a sweetened rice drink. In order to make Sik-hye, cooked rice is cooled down, mixed with water and malt in a container under certain temperature. Malt includes catalyzing enzymes such as α-amylase, β-amylase, β-glucanase. With the help of these enzymes, rice and malt are broken down to glucose and maltose and gives a sweet taste of 'Sik-hye'. This process is an enzymatic process, therefore the container that is used in here is an enzyme bioreactor. Next, the mixture is boiled in high temperature to stop all chemical reactions, and refrigerated. The basic development of bioreactor starts from a smaller scale in a laboratory. The culture system that is used in most bioindustries is an aerobic liquid culture system. Therefore, majority of bioreactor development is focused on the effective aerobic liquid culture.

Let's hypothesize we are trying to develop a new bioreactor to increase the productivity of the desired product. In order to increase the productivity, we need to first design a bioreactor by studying behaviors of liquid, gas, and solid in the bioreactor and then study the fluid mechanics, heat transfer, and mass transfer. An easier way to understand this idea is to think of liquid as the dissolved nutrients in water, gas as air containing oxygen, and solid as the cultured cells. In other words, cells need to absorb appropriate amounts of nutrients and oxygen dissolved in liquid. As cells continuously move, they are constantly contacting with liquid, therefore, nutrients and oxygen are transferred into the cells. Accordingly, constant aeration and mixing are very important. Oxygen in air bubbles is continuously transferred to the liquid by a continuous contact between air bubbles and liquid. As the soluble nutrients and oxygen are supplied into the cells, the cells go through a complex metabolic pathway and produce certain product or by-product and carbon dioxide, and are delivered outside of the cell. Carbon dioxide is delivered back to liquid and comes out as gas form, but all these processes of delivery to liquid and to gas are not possible without continuous mixing. The series of transferring is called mass transfer. In the heat transfer, the entire temperature can be kept constant due to continuous mixing of the liquid in the bioreactor. Therefore, the temperature of the cells and air in the bioreactor is also kept constant. Using this idea, temperatures that rise due to metabolic heat generated by the metabolism of cells can also be kept constant in the same way.

Figure 6-3.
Shake-flask culture in a shaking incubator

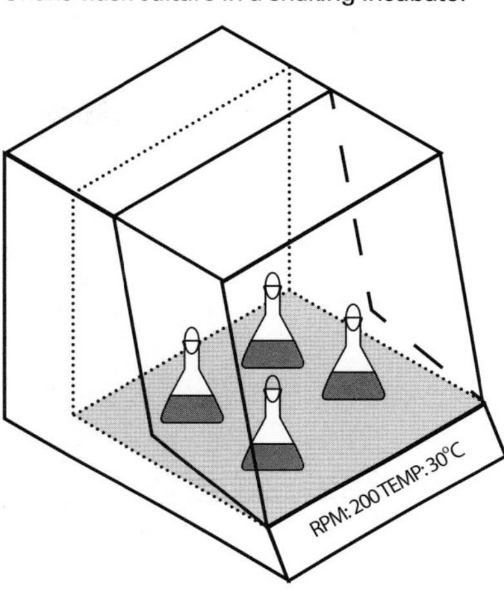

Fluid mechanics is a study of the behavior of fluids in a bioreactor, which can give strong effect on mass transfer, heat transfer, and mixing (Figure 6-3).

New bioreactors can be developed in laboratories through these basic research studies and these types of bioreactors must be scaled up to enable mass production of products in industry. However, large-scaled bioreactors would not obtain the same level of yield and productivity as the bioreactors that are made for laboratory scale. The reason for that is because the fluid dynamics, mass transfer, and heat transfer vary and higher energy consumption occurs as the scale of bioreactors become larger. Therefore, these factors should be taken into account during bioreactor development for improvement of yield and productivity. In general, ensuring that all components that are optimized for the laboratory scale work are properly operating on the medium-sized bioreactors is important during the process of converting from a small bioreactor to a large-scale bioreactor. The bioreactor during this process is called the pilot-plant bioreactor. In this scale of bioreactor, we can check if the optimal conditions that were established in a small-scaled bioreactor can be operated similarly in a medium-sized bioreactor and produce similar level of yield and productivity. When the yield and productivity are below expectations while the similar result

Figure 6-4. Large scale stirred-tank bioreactor

was expected, some modifications are made in some conditions to obtain appropriate level of yield and productivity. Once succeeded, a large-scaled bioreactor is made under these modified conditions and mass production of the product can be performed. If a problem is seen during this process, it is necessary to go back to the small scale and solve the problem. If the modification of a large-scaled bioreactor is desired to produce a different product, we must re-establish starting from obtaining the optimum conditions from a laboratory scale bioreactor to a pilot-plant bioreactor, then bioreactor for mass production. As high value-added products with smaller amounts are preferred nowadays, a pilot-plant bioreactor is often used as a bioreactor for mass production (Figure 6-4).

Classifying different types of bioreactors can often be characterized by different processes of mixing materials – mixing by using mechanical equipments, mixing by sparging air, and by adding a solution including substrates by using a pump.

The bioreactor most commonly used today is called a stirred-tank bioreactor that uses a mechanical device called an agitator or an impeller operated by a motor. With proper and even mechanical stirring and mixing of cells, liquid nutrient medium, and air bubbles in a bioreactor, biological reactions are properly performed. In fact, most of the cells used in the biotechnology industry are aerobic and often require oxygen for biological reactions. Supply of pure oxygen itself is challenging due to its high price,

Figure 6-5. Stirred-tank bioreactor

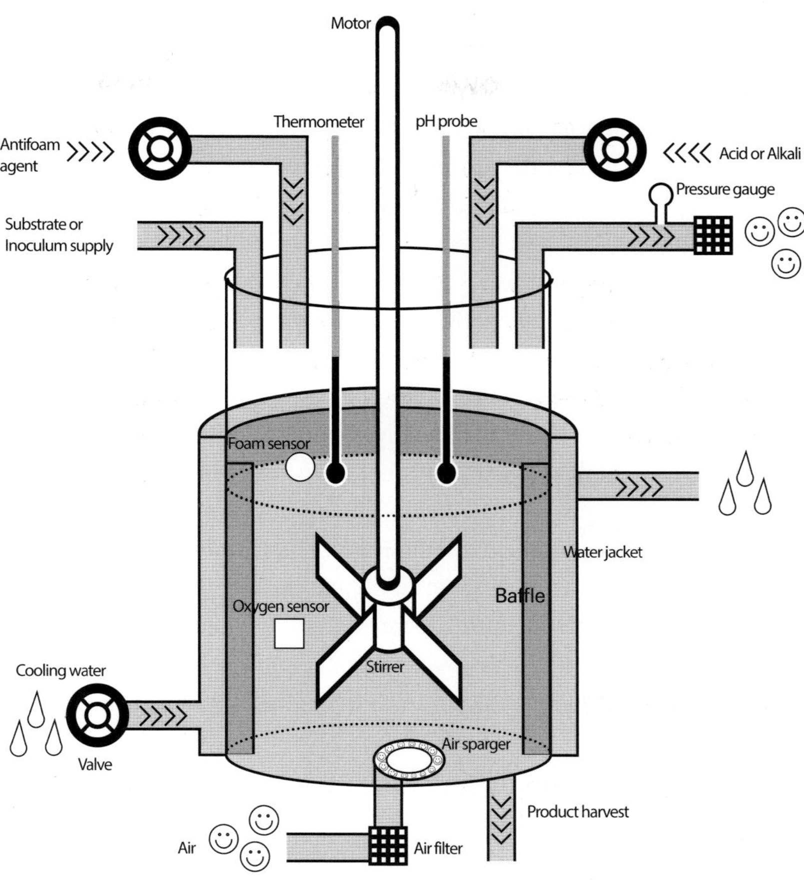

therefore, oxygen-containing air is provided through a device called an air sparger.

The oldest reactor known is probably the reactor that mixes materials with mechanical devices. Consider a machine that is used a lot in our homes, a so-called mixer. A mixer is an impeller that grinds fruits into small pieces and mixes with water. An impeller that is mounted with a mixer is operated by a motor, and the result product changes by adjusting the number of revolutions of the impeller. This type of bioreactor is called a stirred-tank bioreactor, and has many different types such as an impeller that mixes gas, liquid, and solid materials; a baffle that interferes with the fluid flow in the bioreactor to induce more efficient mixing; an air sparger that sprays air for supply of oxygen. Depending on the types of impeller of a bioreactor, the mixing pattern also changes due to a varied flow of fluid. It is the same principle as a propeller in wind power generator or a ship screw.

Therefore, it is important to operate the impeller under the appropriate agitation speed because a shape and speed of an impeller are critical when culturing or reacting cells or enzymes in a bioreactor. Especially when culturing cells, oxygen should be continuously and appropriately supplied by using the air sparger. Generally, the air bubbles that come out through the air sparger are finely broken down by the impeller located directly above and uniformly distributed throughout the whole bioreactor so that oxygen dissolved in liquid is transferred into a cell for metabolism. The smaller the air bubbles, the higher the rate of oxygen transfer to liquid due to the increase of surface area of the air bubbles contacting liquid. As more oxygen is dissolved into liquid, a cellular uptake could be better with higher level of oxygen. If the air bubbles are bigger, the surface area of oxygen to contact with liquid decreases, and oxygen transfer becomes dysfunctional. Therefore, the oxygen transfer rate is critical for efficiently producing high value-added products by using different types of cells (Figure 6-5).

A fluidized-bed bioreactor works differently from a stirred-tank bioreactor and is not mixed by a mechanical stirring device. It is a bioreactor in which only the air sparger is installed, the air goes through the air compression pump and the air sparger, and comes out as small bubbles to be mixed with other fluid. The simplest fluidized-bed bioreactor is made with a cylindrical column with an air sparger located on the very bottom. An example can be easily found in cafes or restaurants – a cylindrical or rectangular column filled with water and fake fish moving up and down by air jetted from the bottom where the air sparger resides. This is exactly how a fluidized-bed bioreactor fundamentally works. This type of bioreactor is called a bubble column bioreactor. This type of bioreactor controls the air flow rate to maintain appropriate mixing. When the seed cells are introduced into a liquid nutrient medium inside a bioreactor and air is injected, this becomes a great bioreactor. There is a slightly modified version of

Figure 6-6. Fluidized-bed bioreactor

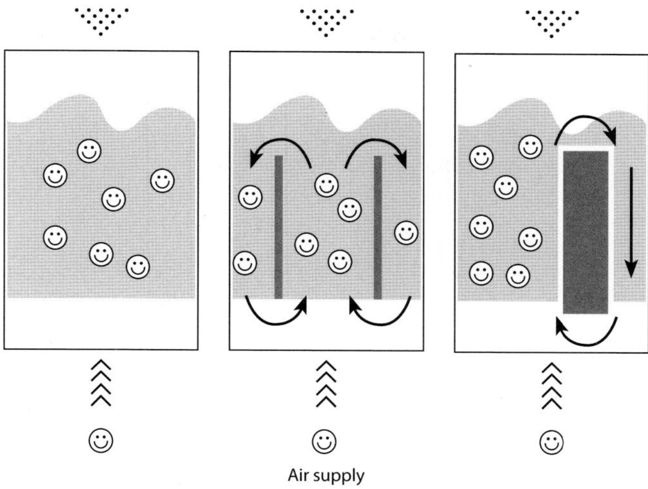

bioreactor called an air-lift bioreactor. This type of bioreactor has a smaller sized cylindrical column inside or outside of the bioreactor that makes the fluid mildly circulated after the air sparging. It can be called either internal or external-loop air-lift bioreactor and these types of bioreactors are mainly used for fungal, animal, and plant cell cultures (Figure 6-6).

A packed-bed bioreactor is often used to convert the solid substrates to another or to perform the bioconversion reactions by enzymes. When a solid substrate is introduced with microorganisms in a bioreactor, the product is produced by a conversion reaction performed by microorganisms. In addition, the bioconversion reaction by the enzymes can produce a desired product if it is carried out by charging the immobilized enzymes into a bioreactor and passing the substrate through the enzymes. The packed-bed bioreactor can make a desired product if the cells or enzymes are charged inside the bubble column-shaped bioreactor that is placed horizontally or vertically, flowing the nutrients of the liquid from one side to the other. Cells or enzymes cannot be charged on their own, so they are entrapped by being packed together inside the solid or semi-solid materials or are chemically bonded together by using the immobilization technique. This bioreactor is particularly used a lot when industrially performing bioconversion reactions using immobilized enzymes (Figure 6-7).

Figure 6-7. Packed-bed bioreactor

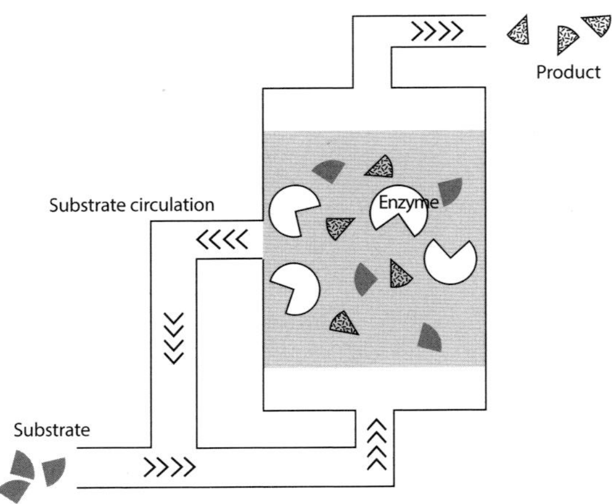

A membrane bioreactor is a bioreactor that uses a membrane with the uniform pores on its surface. Numerous small pores are made through a chemically synthesized thin membrane and only the materials smaller than the pore size can pass through the membrane. For example, there is a typical membrane bioreactor with multiple long and hollow fibers inserted into a cylindrical module. The hollow-fiber membrane is made as a hollow tube that is deep and thin. A bundle of hollow-fiber is placed into a certain sized cylindrical tube and is operated. Charging the cells or enzymes at the empty space inside the hollow-fiber membrane and finishing the both ends well with the hollow-fiber membranes placed inside of the cylindrical tube completes a good bioreactor. As a substrate is supplied from one side as in a packed-bed bioreactor, the substrate reacts with the cells or enzymes to produce a product. When it comes out through the membrane, it can be collected from the other side of the end (Figure 6-8).

As the examples mentioned above, various types of bioreactors exist, and an appropriate bioreactor should be selected or designed according to the system or cellular characteristics. Ultimately, this is aimed not only in mass production but also in improving the final productivity and yield of the product. Yield and productivity are critical factors with its close relation to economic efficiency. Yield may vary depending on the type of the end product, but generally refers to the ratio of the amount of the final

Figure 6-8. Membrane bioreactor

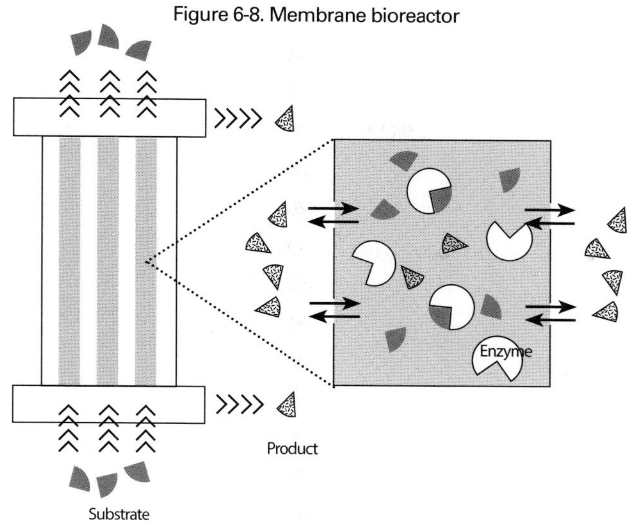

product produced with the basis of the amount of nutrients consumed. Also, productivity refers to the concentration of the final product produced under the certain period of time. Besides, appropriate selections of cells or enzymes should be made depending on the characteristics, and the system should be established by examining the reaction characteristics under the optimum culture or reaction conditions at the laboratory scale. Although many industries prefer to choose the suitable industrial bioreactor that is already established and that fits into the desired system, these industries still need to design their own bioreactors that is necessary for their system or desired product. In addition, there are still many possibilities to develop a bioreactor according to the purpose of the system.

The proper selection and operating conditions of the bioreactor that fits the characteristics of microorganisms or animal or plant cells have a great influence on the productivity of the whole process. For example, the mechanical impeller in a stirred-tank bioreactor may damage some fungi, plant, or animal cells. This is a reason why a special impeller is developed or a fluidized-bed bioreactor is used. Cells and nutrients inside the bioreactor are properly mixed under the optimal culture conditions. These nutrients such as carbon, nitrogen, or inorganic salt sources should be supplied properly, and pH, temperature, and dissolved oxygen should be measured and controlled for efficient cell growth and metabolism. The most important factor is the continuous supply of oxygen at a constant concentration that is applied under all aerobic cultivation processes. All of the oxygen concentration, oxygen uptake rate, and oxygen transfer rate should be at the optimum level. A problem regards to this is that the concentration or metabolic activity of a cell may change due to the external environment, so it may be difficult to balance the supply and demand of oxygen. In most cases, the supply limitation is higher than the demand, so one good way is to decrease the demand of oxygen to a microorganism without impacting towards productivity. The necessity of modification or designing new types of bioreactor increases as the cell and the useful products vary in wide ranges. Thus, it is very important to select or develop a bioreactor suitable for a biological system.

Chapter 7

Application of Cells and Enzymes for Efficient Production of Biotechnological Products

Before the development of technologies that utilize cells or enzymes in biotechnology field, most of the desired substances were synthesized through chemical reactions. However, as mentioned in Chapter 4, chemical reactions generally have complicated processes due to the use of catalysts and organic solvents at high temperatures and pressures. Also, too much energy is needed to perform these reactions and there is a higher risk of environmental contamination. Since many byproducts can be produced along with desired products, many attempts have been made lately to produce desired product that are eco-friendly under normal temperature and pressure (Figure 7-1)

In order to produce various biotechnological products by using cells and enzymes mentioned in Chapter 3 and 4, biological processes from Chapter 5 should be economically designed, and the bioreactors mentioned in Chapter 6 need to be made with its functional efficiency. For the application of the cell, it is necessary to have a genetic stability during high cell density culture in a bioreactor. Also in the case of enzyme application, it is necessary to consider the economical view for bioconversion process in an enzyme bioreactor due to its high cost in enzyme production and partial or complete purification of enzymes. During cell culture, even though some portions of cells come out of the bioreactor or are dead, remaining cells can still grow constantly and have consistent concentration by using the remaining nutrients. Moreover, product yield and productivity can be increased with higher cell concentration when immobilization technique is used due to more porous and larger surface area of a carrier. To give an

Figure 7-1. Chemical reaction and biological reaction

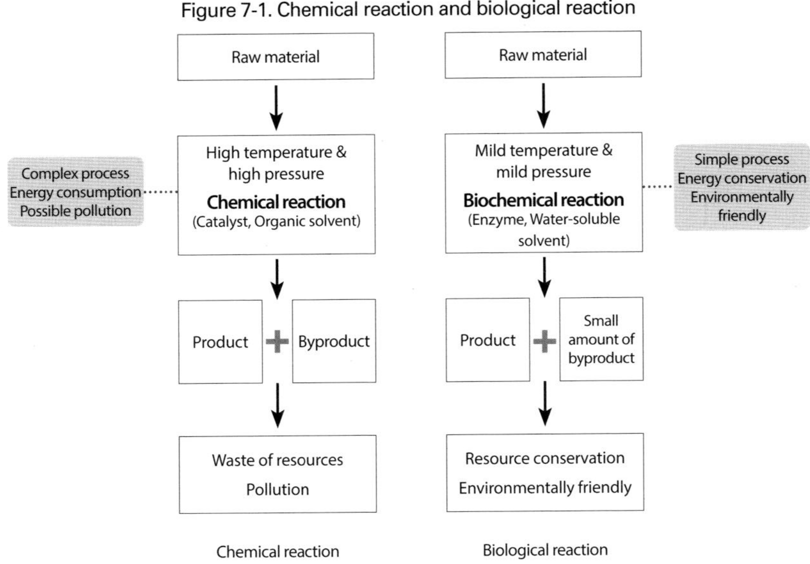

easier understanding, dish washing sponges tend to have small pores all around for higher surface area. These types of sponges can be made in different sizes and shapes. Materials that can be used to produce a carrier include various organic and inorganic materials and nano-sized materials which can be made with either organic or inorganic materials (Figure 7-2). On the other hand, enzymes cannot be used for bioconversion if they are partly denatured or come out of the bioreactor. Therefore, when free enzymes are used during bioconversion process, only one single reaction is allowed and considered as 'economically poor'. For industrial application of enzymes with higher economical value, enzymes can either be bound to a solid or semi-solid materials or be entrapped. Then the immobilized enzymes can be used for bioconversion continuously or a couple of hundred times. This bioconversion process could relatively be economical, compared with chemical synthetic reaction in general. Thus, various enzymes can be applicable for the efficient industrial application based on the immobilization techniques.

Immobilization technique is a technology of chemically adsorbing, binding or entrapping the cells or the enzymes to various solid or semi-solid

Figure 7-2. Various carriers for immobilization

	Property	Method
Graphite	A crystalline structure of the carbon with its atoms arranged in a hexagonal form.	Adsorption, Covalent bond, Crosslinking, Ionic bond
Alginate	A natural biopolymer which can be extracted from brown seaweeds.	Entrapment
Sillica	The common name is silicon dioxide (SiO_2)	Covalent bond, Crosslinking, Ionic bond
Alumina	It is also called aluminum oxide (Al_2O_3)	Adsorption, Covalent bond, Crosslinking, Ionic bond
Zeolite	An aluminosilicate mineral.	Adsorption, Covalent bond, Crosslinking, Ionic bond
Chitosan	A polysaccharide produced by deacetylation of chitin.	Entrapment
Acrivated carbon	A form of carbon which has a high surface area.	Adsorption, Covalent bond, Crosslinking, Ionic bond

materials called carriers. Physical methods include the cell immobilization that limits the free-moving of cells of microorganisms, plants, or animals, and the enzyme immobilization that is used to physically or chemically entrap or bind enzymes into a carrier for bioconversion of one material to another. In addition, when large amount of desired enzymes are produced in a cell through genetic manipulation, it is possible to give a similar effect to that of immobilizing enzymes within the cell. Displaying much desired enzymes on the surface of *E. coli* or yeast cell called the surface display is also being actively developed. All these types of technologies are called whole cell immobilization. Immobilization techniques are widely used in many different industries because it can increase the productivity in bioprocesses using cells or enzymes (Figure 7-3).

The advantages of using an immobilized cell is that there are less washouts of the cell that escapes outside the bioreactor by immobilizing the cell into a carrier, therefore, the desired product can be continuously produced under low contamination environment. Also, bioreactors can be easily used with low cost while maintaining high concentration of cells by immobilizing them into the carrier. Some disadvantages exist, too. In general, cells located inside or outside of the layers of carriers may not

Figure 7-3. Typical immobilization methods

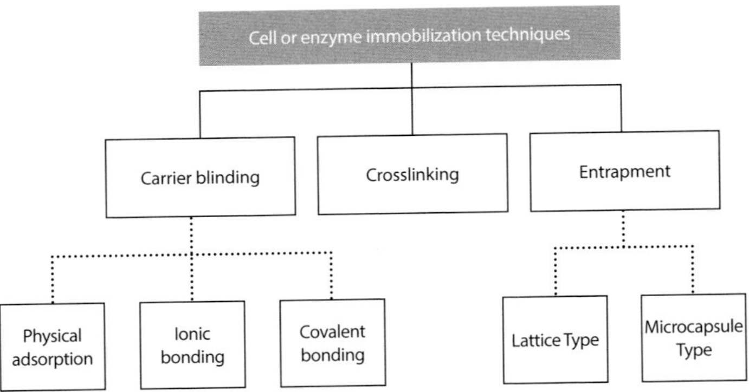

be able to properly uptake oxygen and nutrients, resulting in partial loss of cell activity and survival. However, this partial loss can be overcome by having high initial cell concentration inside the carrier. Even then, the high concentration of cells inside the carrier can also bring a problem in mass transfer by disabling the transfer of materials through the carrier. In other words, not enough nutrients are being transferred over to the center of the carrier for cell growth, and these cells start to change the metabolism to produce different metabolic by-products. Accumulation of these unnecessary by-products decreases the activity of the cell, and the overall productivity decreases at the end. Also, the calcium alginate gel, which is the widely used material for carriers, can easily be destroyed in the presence of components such as phosphate. Even though these disadvantages exist, the advantage of cell immobilization has been utilized for the industrial applications. As examples, production of ethanol, organic acids and antibiotics using microorganisms, or production of useful metabolites using plant cells, or production of monoclonal antibodies using animal cells are all popular fields that are currently under active research by using different types of cell carriers and producing more industrialized products continuously.

Though many different methods of cell immobilization have been carried out, the attachment and entrapment methods are most widely used as the current cell immobilization methods. For attachment method, the cells could be adhered onto the surface of a carrier by using two different types of forces such as an adsorption method that includes static electricity, ionic bonding, or natural biochemical affinity, or a covalent bond method that uses chemical bonding. Adsorption method has some advantages that are simple, cheaper, and less activity loss of the cell due to the use of mild reaction conditions. However, this method also has some weaknesses such as the cell can be easily detached from the carrier after adsorption due to pH or salt concentration. The covalent bond method is used more widely for enzyme immobilization rather than cell immobilization. The functional groups such as esters, bisaldehydes, and amine groups are chemically bonded with the carrier. Other than these two methods mentioned above, crosslinking method is also used to link the enzymes by using covalent bonding. By using this method, the crosslinking agent that has two or more of the same types of the functional groups can be used to link same or different types of enzymes. One of the most well-known cross-linking agent is glutaraldehyde that has same type of functional group on each side. All these methods can be easily handled, are easily reused by simple recovery through filtration or centrifugation, and the reaction can be stopped at any specific stages. We will talk more about enzyme immobilization method later.

The entrapment method is confining the cells into the carrier which is possible to diffuse nutrients and products. To make a gel to confine the cells, synthetic materials such as polyacrylamide, polyurethane, or natural materials such as collagen, alginate, carrageenan, or agar are much used as the gel material. Though synthetic materials make more stabilized gels, they cannot be manufactured under the mild conditions as the natural materials. Therefore, the entrapment method by using natural polymers is most popularly used (Figure 7-4).

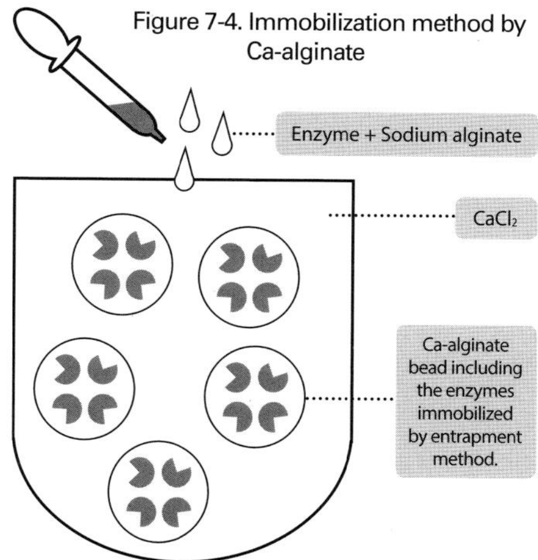

Figure 7-4. Immobilization method by Ca-alginate

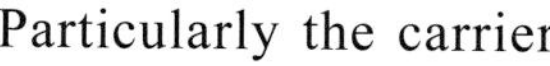
Particularly the carrier

used to immobilize cells has a big influence on cell activity and efficiency of fermentation process, therefore, many factors must be carefully considered and chosen. Cell activity and the mechanical strength must be high, carrier materials must be able to be found easily in mass-scale, immobilization cost must be low, and manufacturing process must be easier. Considering these factors, agar has low mechanical strength, carrageenan has problems in becoming large-scale. Therefore, alginate is the popular option in laboratory scales for research, and is being used for industrialization. The most popular carrier material, alginate, is mainly produced from seaweed materials but especially produced from *Macrocystis pyrifera*.

Currently, most of the transportation fuel ethanol produced worldwide in the field of bioenergy is produced by batch culture. In this culture, the processes such as washing, sterilization, filling, inoculation, fermentation, and recovery of the fermentation liquid are repeated. Though, this repeated procedure has high labor cost and difficulty of automatic control. If continuous fermentation using immobilized cells is used, productivity can be improved and overcome the limitations mentioned above.

Let's talk more in depth about the enzyme immobilization now. Consider a simple process of enzymatic reaction using sucrase that breaks down sucrose into glucose and fructose. When sucrose is bound to the active site of sucrase, a certain reaction is carried out and produce glucose and fructose. Then, all sucrose, glucose, fructose, and sucrase are mixed together inside an enzyme bioreactor which makes separation of the products difficult, and limits rest of the sucrose to keep reacting with sucrase. This is why chemically binding or physically immobilizing sucrase with solid or semi-solid substance can make a recovery process of desired product much easier, while keeping sucrase to continuously react with sucrose to produce more glucose and fructose.

Several advantages can be addressed if enzymes are immobilized in this way. The enzymes can be used repeatedly, the stability is improved, the reaction can be easily stopped, and the product can be easily separated. Also, various types of enzymes can be separately immobilized in the carrier or can be immobilized all at once to perform the reaction.

The most important part to consider when using an immobilized enzyme is to fully understand its biochemical characteristics, reaction

Figure 7-5. The important factors for enzyme immobilization

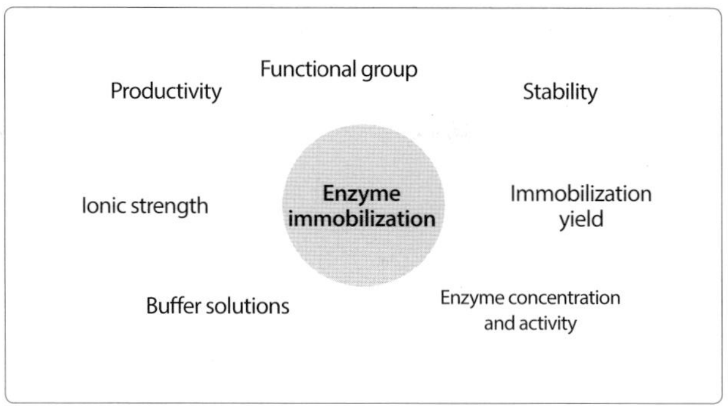

mechanisms, and chemical and mechanical properties of the carrier. By focusing the basis on these points, an appropriate immobilization method should be chosen and proper performance of mass transfer should be made sure. Not only that, it is also important to examine whether the stability of the process is reliable, and conversion rate of the substrate should be evaluated at the final stage when operating in an enzyme bioreactor. The biochemical characteristics of an enzyme include molecular weight, functional groups on surface of an enzyme, and purity. It is also important to know the specific activity, a parameter of an enzyme mechanism, effect of acidity and temperature, inhibition by an inhibitor, and stability in the presence of different substances. The chemical property of a carrier includes chemical components, functional groups, size, and stability. Lastly, the mechanical property refers to the strength of the carrier when various enzyme bioreactors are used.

To select a suitable immobilization method, evaluation on how much the enzyme proteins are attached or immobilized on the surface of a carrier, and clear calculation on the active enzyme yield are necessary. It is also crucial to know the type and concentration of the buffer solution, and to examine if the substrate or product diffusion is well-facilitated. When it comes to stability, it is important to examine whether the stability is maintained during the operation of enzyme bioreactor and when the immobilized enzyme is kept frozen for a long time. Eventually, the conversion rate and productivity should be evaluated to determine the economic viability (Figure 7-5).

Figure 7-6. Immobilization by a cross-linking agent(Glutaraldehyde)

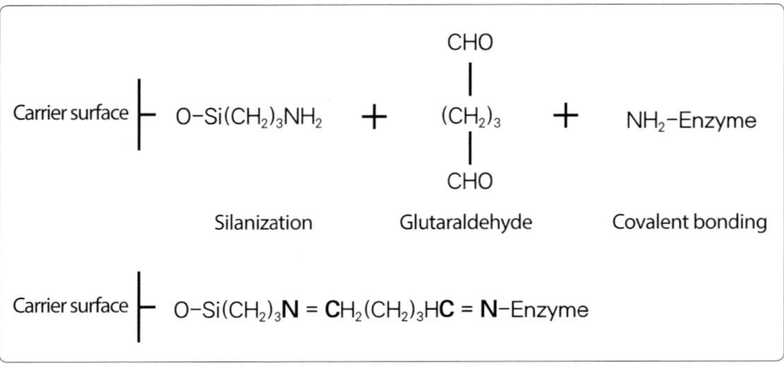

Normally a method using the crosslinking agent is mostly used to immobilize an enzyme to produce a certain product. An enzyme is either linked to the surface of a carrier by using the crosslinking agent, based on covalent bonding or enzymes themselves are linked together without a carrier with the use of the crosslinking agent. The reason why covalent bonding is most popularly used is because the enzyme must be attached firmly onto the surface of the carrier during the conversion reaction in order for the reaction to perform without a problem. Many types of functional groups such as an amino group, a carboxyl group, a hydroxyl group, and an imidazole group are mostly used for covalent bonding. When using these functional groups, the reagents used should not affect the active site of the enzyme, and the enzyme must maintain its activity. However, there are possibilities of the active site to be affected during the reaction. Since the reaction process is quite complicated and quite costly, it is very important to consider and evaluate the above facts and possibilities carefully (Figure 7-6).

A comprehensive and detailed technology is necessary in order to apply the enzyme immobilization technique in industries. The economic strength of an enzyme immobilization on a carrier only shows when the reaction brings out a high yield and productivity. We will talk about the detail of enzyme immobilization process based on the covalent bonding. First of all, removal of all impurities on the surface of a carrier should be needed, and the functional groups should be created for the enzyme or the crosslinking agent to be linked to the surface of a carrier. Amino groups

Figure 7-7. Enzyme immobilization by pretreatment method

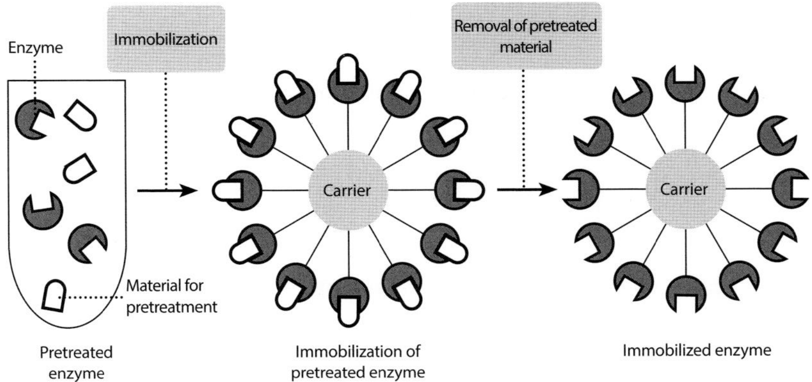

are often produced as a functional group, and it is necessary to have an optimization process with the suitable reagents because the reactions are influenced by acidity, concentration, or reaction time.

When the enzyme is directly linked to the amino group formed on the surface of a carrier, not many enzymes are attached to the surface due to the size of the enzyme. Therefore, enough space should be saved for the enzymes to attach by using a crosslinking agent called glutaraldehyde that has a carboxyl group on both ends. By linking the glutaraldehyde with the amino group, it creates more space for the enzymes to be attached on the surface. Glutaraldehyde may affect in the enzyme activity, so it is also used after being chemically modified. During this process, the optimization process is necessary because the acidity, temperature, time, and concentration of glutaraldehyde may affect the reaction.

The enzyme can now be immobilized by linking the amino group of an enzyme to the carboxyl group on the other side of the glutaraldehyde. Concentration of enzymes, types and concentration of the buffer solution, reaction time and temperature can affect the reaction during this process. At the end, fully understanding the characteristics and stabilization of the immobilized enzyme for reuse evaluation is important. There are times when the activity of the enzyme is lost during the immobilization process because the crosslinking agent interrupted the active site of an enzyme. To prevent this possibility, an appropriate substrate is bound to the active

Figure 7-8. Stirred-tank bioreactor with the immobilized enzymes

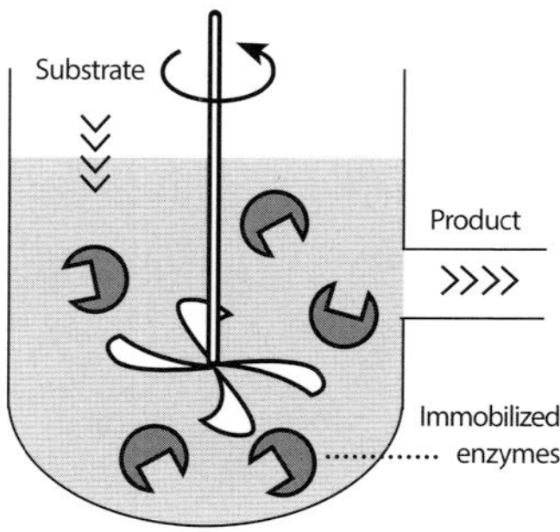

site of an enzyme before immobilization and then link the enzyme to the carrier. This increases the activity and stability of an enzyme by forcing the crosslinking agent to be attached to the non-active site. This type of method can also be applied to various enzyme conversion reactions (Figure 7-7).

As an immobilized bioreactor using enzymes, a stirred-tank bioreactor and a packed-bed bioreactor are mostly used. As an appropriate amount of immobilized enzymes are added into the bioreactor with addition of the substrate fed at a constant rate, the immobilized enzyme and the substrate react together for a certain period of time, and the product made from the reaction can be continuously collected. Because conversion rate is very important, the feeding rate of substrate and the concentration of immobilized enzyme should be controlled properly. If the reaction rate and conversion rate decrease, the product solution can be recycled or the bioreactor can be connected in two stages for the two bioreactors to continuously circulate the substrate solution (Figure 7-8).

As an example of an industrialized process mentioned above, a packed-bed bioreactor is used when glucose is converted to fructose with the enzyme immobilization technique. An enzyme called glucose isomerase converts glucose to fructose. When glucose isomerase is immobilized into an appropriate carrier, added to the packed-bed bioreactor, and glucose solution is added to the bioreactor as a substrate, fructose can be continually collected as a desired product. As time goes by during this process and as the activity of the immobilized enzyme decreases, a different bioreactor that is charged with new immobilized enzyme is used while the old bioreactor with decreased enzyme activity can be regenerated and recycled for the later use.

Production of fructooligosaccharides has been industrialized based on the whole cell immobilization which is actively under research for various bioprocesses in industry. Making the yeast to produce large number of transferases that can convert sugars into fructooligosaccharides, immobilizing the yeast into the carrier, and charging it into a stirred-tank or packed-bed bioreactor bring the whole-cell immobilization as an industrialized process. When sugar is fed into the immobilized bioreactor, it is broken down and becomes glucose and fructose, and these fragments chemically bind together in random orders to become oligosaccharides. Oligosacchardes have the sweetness when ingested, but cannot be used as an energy source in a human body due to the lack of enzymes. Instead, oligosaccharides are ingested and used by intestinal microorganisms when they arrive in the intestines.

Let's look at an example of some current researches on immobilization by surface display for the efficient production of bioethanol and biodiesel. For bioethanol production, cellulosic material is hydrolyzed to glucose by cellulases, then glucose is fermented by yeast. This process normally consists of two steps, but can be carried out in one step if the strain is developed by surface display. If yeast is genetically manipulated to express cellulases on its surface, the surface-expressed cellulases can break down cellulosic material into glucose and the glucose is fermented to bioethanol by yeast. In addition, bioethanol can also be continuously produced at a high concentration if the enzyme-expressing yeast on surface is immobilized on the carrier.

To produce biodiesel, vegetable oils are generally used as the substrates such as soybean oil, palm oil, canola oil, or jatropha oil that are mainly made up of triglycerides bonded with various fatty acids. As ethanol or methanol are added, the substrates go through hydrolysis and esterification

Figure 7-9. Immobilization by surface display method

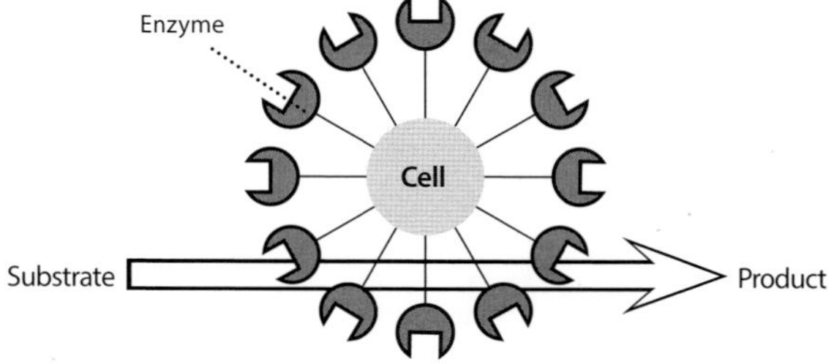

with lipase, and the biodiesel is produced in the form of either fatty acid methylester or fatty acid ethylester. During this process, biodiesel can be continuously produced if lipase is immobilized in a carrier and added to the bioreactor with the constant feed rate of oil and a proper amount of water. Surface display studies allow yeast or *E. coli* to express lipase by gene manipulation, and can be performed in a packed bed bioreactor after immobilization of the cells.

The immobilization technique using surface display could be applicable to biosensor-related researches that can be used in expressing proteins such as antibodies and viruses, and diagnosing diseases or environments are currently developing actively (Figure 7-9).

β-Lactam antibiotics such as penicillin or cephalosporin cannot be used on their own due to the antibiotic resistance to pathogens. Therefore, cells are cultured in a bioreactor and the antibiotics are produced. After that, the chemical structure is modified by immobilization technique, and is used as antibiotic intermediates. The global market for these antibiotic intermediates is quite large, and they are constantly making different types of antibiotics that can fight against various pathogens by changing structure with the help of chemical synthesis, cells, or enzymes (Figure 7-10).

As a result, the overall productivity can be increased by high cell density culture or conversion in a bioreactor that is selected under the suitable immobilization technique. Immobilized bioreactors are critical for producing useful biotechnological products along with the gene recombination and cell fusion technology, and can be used in a broad spectrum. Development of immobilization technique in a bioreactor has greatly influenced in various bioindustries such as pharmaceuticals, food, chemicals, agriculture, and environment. Now, let's look deeper into the

Figure 7-10. Production of Antibiotic intermediate by immobilized enzyme

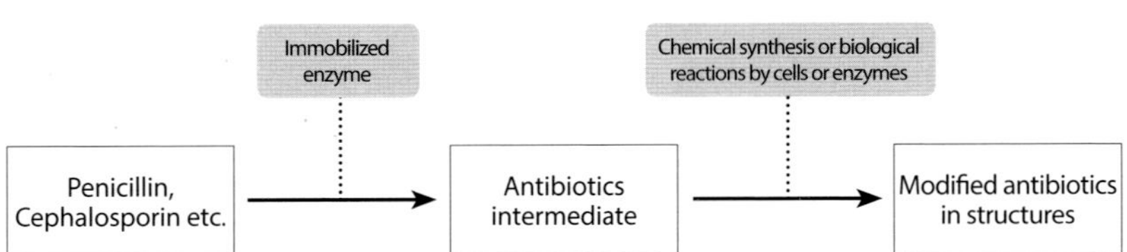

Figure7-11. Application of immobilized bioreactors

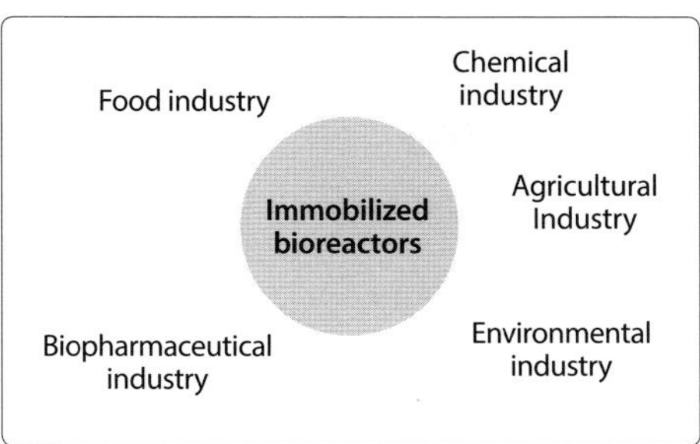

application of these bioindustry fields (Figure 7-11).

The use of bioreactors in the biophamaceutical field began in 1944 with the production of penicillin after establishing mass culture technology of *Penicillium* species. In the early period of penicillin production, the concentration of penicillin was too low. After establishing the right strain by screening and mutation, and the optimum culture condition, the production of penicillin has increased. This has built strong basis of the bioreactor and has been applied for the production of various bioproducts produced by useful organisms such as microorganisms, animal cells, or plant cells. As the immobilization technique has developed, *Penicillium* species can be immobilized in a carrier that enables penicillin to be semi-continuously produced which increases the overall productivity. The biopharmaceutical products that can be attempted for mass production by immobilized bioreactors are hormones (insulin, human growth hormone), antibiotics (tetracycline, cephalosporin), anticancer drugs (interferon, interleukin, tumor necrosis factor), monoclonal antibodies, thrombolytic agents (urokinase), erythroid promoters (erythropoietin, colony stimulating factors), and crude drugs (berberine, ginseng).

Immobilization technology is also widely used in drug delivery system. Immobilization technology is applied in encapsulation of drugs to prevent the drugs from being destroyed by gastric acid when administered into the body. As a material for encapsulation, biopolymer materials are used so that the drug can be slowly introduced inside the body at small intestines by slowly releasing the drug from the capsule without being destroyed in the stomach.

The most representative food products include sweeteners (isomerized sugar, aspartame, fructooligosaccharides), seasonings (glutamic acid, lysine), and fermented foods (beer, sake, soy sauce, vinegar). Among these products, isomerized sugar is obtained during the conversion process of glucose to fructose by the enzyme glucose isomerase that is secreted by bacteria. Fructose is commercially available as a syrup product because it has about 1.6 times stronger sweetness than that of sugar at low temperature. With the availability of continuous production operation of the immobilized isomerase enzyme inside the packed-bed bioreactor, millions of tons of fructose are being produced worldwide. Another example would be amino acids. About 30% of the total global market production of glutamic acid and lysine are currently being supplied from South Korea, and these amino acids are mainly used as an additive for seasonings, sweeteners, and animal feeds. Much research has already been conducted on the production of these amino acids with the use of immobilized bioreactors.

There are some interesting pigments and spices used as the ingredient of cosmetic products in the chemical industry. Pigments extracted from plants include β-carotene, shikonin, and chlorophyll. Shikonin is a type of pigment that is obtained from the roots of plants. With the success of mass production of shikonin in a bioreactor as the technology of plant cell cultivation developed, it is widely used as the raw material of lipstick. More researches are being done currently with application of immobilization technique. Production of alkaloids by plant cell cultivation is mainly performed in a fluidized-bed bioreactor.

As a representative of applied agriculture, immobilization technique is used to develop microbial pesticides as biological pesticides. Especially, spores of entomopathogenic fungi are being developed in multiple ways by using the principle of these spores infecting and annihilating pests. Entomopathogenic fungi look different from mushrooms, but has very similar life cycle. These fungi parasitize the body of an insect during the winter, and grows out of the body as a thread-like plant by using the nutrients of the body. When it starts growing inside the body, it branches out the vegetative part consisting a network of fine white filaments called hyphae. A group of hyphae is called a mycelium. When the temperature and humidity are optimum during the summer, a mycelium fully colonizes

Figure 7-12. *Cordyceps nutans Pat.* (Left) and *Cordyceps sphecocephala* (Right)

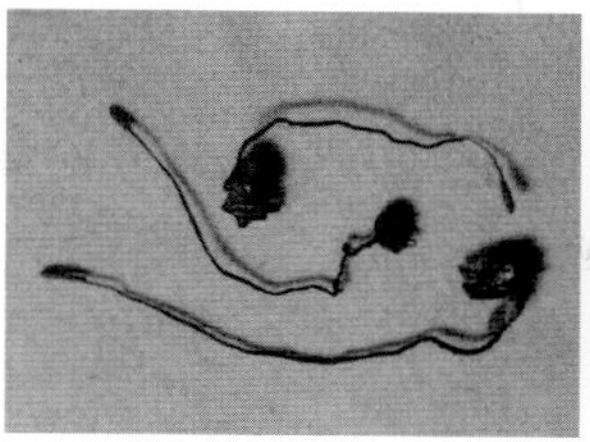
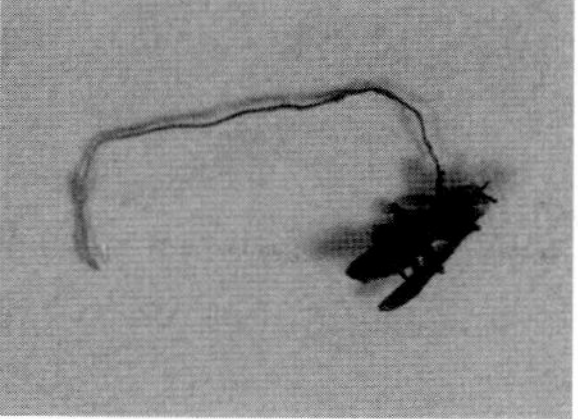

and start to grow a fruiting body that looks like the body of a mushroom. Top of the fruiting body normally reproduces spores. These spores can easily be transferred by wind into an insect's body, and start growing hyphae again (Figure 7-12).

Immobilization technique is used to formulate spores after mass production and recovery of those from a bioreactor. Using the immobilized spores on plant farms and greenhouses reduces the use of synthetic pesticides, which is better for the environment. Spores are often stored inside of carriers for stabilization. When the spores are mixed with nutrients and sunscreen agents and immobilized, this mixture can be sprayed to the plants to kill pests. Currently, the mixture of synthetic and microbial pesticides would be used.

The environment field is currently being recognized as a very important area due to the ability of solving some environmental problems with a bioreactor. Problems regard to water pollution prevention, waste treatment, marine pollution treatment, clean technology and production of environmental products can be well-solved. In particular, production of biodegradable plastics and pollution-free pesticides, the decomposition of refractory materials, and the removal of pungent odors are currently actively under research. Much has already developed in the research field of wastewater treatment and odor removal processes using immobilized bioreactors.

Chapter 8

Basic Concept for Bioseparation and Formulation of Biotechnological Products

Once I was watching an older Korean drama that was made with the background of when Korea was still the Joseon dynasty, the main character simply grew fungi, obtained what is thought to be antibiotics from it, and gave an intravascular injection to a patient to treat a disease. Seeing that part, I remember getting quite worried about the patient's death because of a bad side effect. Obtaining antibiotics from a fungal culture may have been a groundbreaking technique back then, but using partially separated and purified antibiotics is very dangerous. As shown, separation and purification is such an important portion in biotechnology field.

Chapters 6 and 7 described the mass production of cells for obtaining higher quantity of products and the bioreactors, and immobilization technique for bioconversion of one substance to another by using enzymes. Though desired product is produced in a bioreactor, it is necessary to separate the pure product from other different substances from now on. Chapter 5 briefly talked about this concept, but Chapter 8 will go more in depth with detailed concept of bioseparation. In here, different substances may be the cell itself, unreacted nutrients, or by-products. Thus, the techniques required for product separation or the easiness of separation depend on the characteristics of the final product. As mentioned before about different types of products, for example, a final product could be either a cell itself or the metabolites. If the product is the cell itself, only the cell can simply be separated from the liquid medium that was included in the bioreactor. If the product is metabolites, however, an active strain needs to be used to produce a desired product. The separation process is quite complicated

Figure 8-1. Different types of product in bioseparation process

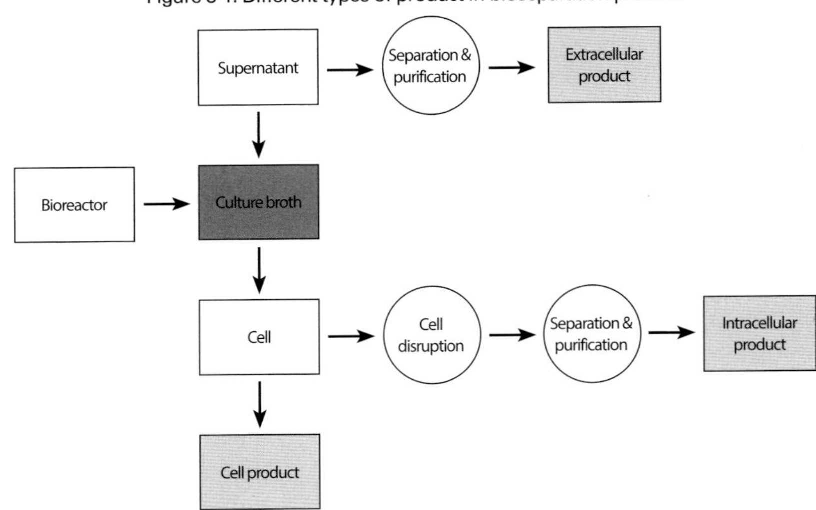

because most of the product is produced into the liquid medium and is mixed with other by-products and various substances that located inside the liquid medium. Therefore in this case, the cell must be separated first, then the desired substances need to be separated from various substances from the liquid medium. Separating only the desired substance is not as easy as it sounds because the products produced from the cell are normally present in a low concentration in the aqueous solution. Some products even accumulate inside of the cell. In general, accumulation can be seen during production of recombinant proteins through gene recombination. In this case, a cell must be separated first, and the desired product accumulated inside the cell are recovered. To perform this, a general method is to disrupt the cell wall or cell membrane, then the intracellular product can be recovered. As shown, the difficulty of separation and the required technique could be different according to the characteristics, purpose, and price of the various products. As the final product is collected from microorganisms, plant cells, and animal cells, different type of process is chosen respectively, and different type of design is needed for efficient bioseparation depending on the physical and chemical characteristics of the final product and its by-products (Figure 8-1).

Figure 8-2. The important factors for optimal design of bioseparation

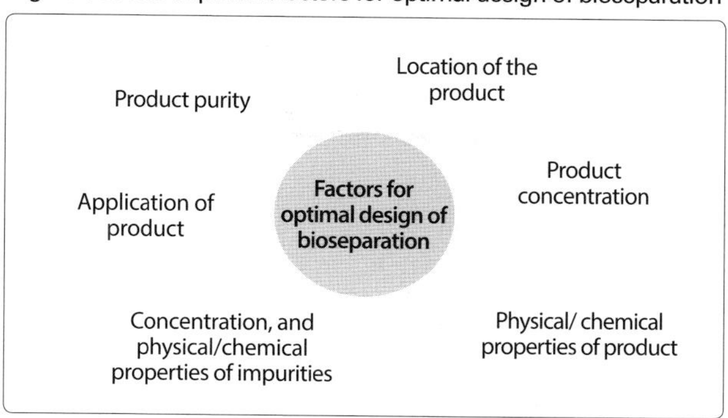

The basic principles of bioseparation are originated from chemical or biochemical laboratories and chemical processes to separate various materials. As many separation methods have newly developed and have gone multiple modifications for a long time, the suitable methods for biochemical products have had established. This chapter will include rather easier concept of bioseparation than processes.

The most important part in using bioseparation is to choose the process with low cost and the fewest steps to get the maximum yield and productivity. The general character of biotechnological products is that the final product normally exists in a very low concentration inside the bioreactor, therefore, selection of the right method for efficient separation is critical. The final product collected from the bioreactor can also be easily decomposed by external environmental condition or through reaction with other by-products. Therefore, these final products must go through fast bioseparation process under the optimum temperature and pH (Figure 8-2).

When bioseparation is performed in an industrial scale, both purification and yield efficiency decrease because it takes a long time for the separation of the final product due to the large size of a bioreactor. As the size is bigger, larger amount of solvents and substances are needed for separation, and the cost also increases. In order to reduce the overall cost, we have to consider reusing the remained substrate that has not been consumed in the bioreactor. Environmentally, the waste and wastewater should be treated economically.

As mentioned above, the basic concept of the product separation focuses on either separating the solid (cell) and the liquid (filtrate without cell) or separating the product dissolved within the liquid. How much purity percentage should be acquired through this separation process strongly depends on the purpose and use of the product. Since the cost difference is mainly depending on the purity of the final product, each step of separation and how much steps are needed in the process should be carefully considered. Therefore, minimizing the number of subsequent processes and choosing rather cost-efficient process is very important. Especially, the bioseparation processes may cause higher damage to the final product due to more complex and more steps required than the chemical separation processes, therefore, it is important to put much more attention during bioseparation.

Let's look into the methods of cell separation, extracellular product separation from the liquid medium, and intracellular product separation from the cell. One of the Korean traditional method of separating yeast was used during the Korean rice wine making process. As the rice wine is made, yeast was easily separated when it was settled on the very bottom of the jar due to gravitation or when it floated on top of the liquid surface due to the buoyancy. This method is still widely used in the process of brewing or wastewater treatment. The most common methods of cell separation are filtration and centrifugation. Before using these methods, a pretreatment method called the broth conditioning is used to aggregate cells within the medium for more effective cell separation. The simplest filtration is a method of separating the solids and liquids by applying pressure using a round filter paper in the laboratory. If the solid is a cell, the size of the cell desired to be filtered should be larger than the diameter of the pore of the filter paper for successful filtration. Filtration can separate materials by pressure difference based on the size of different particles. Since long time ago, filtration has been used to obtain clean water and it is a very important process in brewing technology. The critical factors influencing filtration are the size, shape, viscosity, density, solids content, and scale of the material desired to be separated. The most popular filters used in filtration are the filter press and the rotary drum filter. The current yeast type used in brewing or baking is sold as a cream type after being separated through filtration or in a powder type after dehydration. Most of the filtration are performed in large-scale devices.

Centrifugation is a system for separating materials by the density difference between solids and liquids, and the apparatus used for centrifugation is called a centrifuge. A centrifuge is generally used when the use of filtration system is not feasible. The very first centrifuge was developed by a Swedish engineer, Gustaf de Laval (1845-1913), for the purpose of separating milk cream. Since then, the centrifuge has consistently been developing. The present centrifuges can be divided into low-speed centrifuges, high-speed centrifuges, and ultracentrifuges depending on the rotation speed. The ultracentrifuge is especially used a lot in biotechnology research fields because of its ability to separate various cellular materials including the colloid solution. The ultracentrifuge was first developed by a Swedish chemist named Theodor Svedberg (1884-1971) who is known for his work in separating high molecular substance such as protein and measuring its molecular weight. He was awarded with the Nobel Prize in 1926 for his research on the dispersion system and development of the ultracentrifuge. For reference, the svedberg (S) unit means 10^{-13} seconds in sedimentation constants of colloidal particles. The Uppsala University in Sweden has a research institute called the Svedberg Laboratory (TSL) that was named after him. Ultracentrifugation is known to be more expensive than the filtration system, but it is a faster and more efficient system (Figure 8-3).

Figure 8-3. (a)-Filtration units and (b)-centrifuges

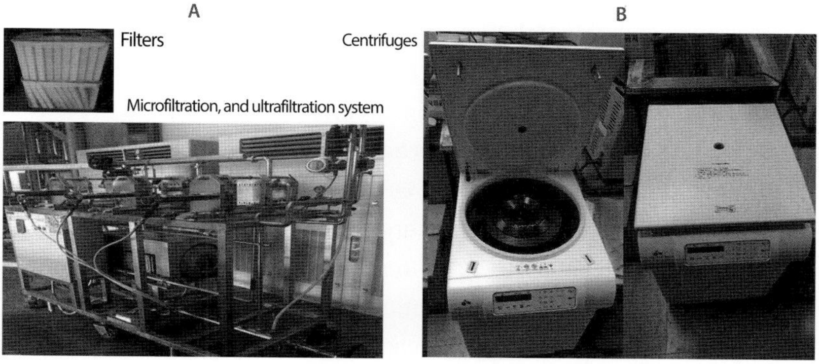

A

B

Filters

Centrifuges

Microfiltration, and ultrafiltration system

Figure 8-4. Distillation equipment

To acquire a desired product from a liquid medium, a cell must be separated first and other desired metabolites should be separated from the liquid medium. Normally, a cell releases metabolic substances out to the liquid medium. Ethanol, amino acids, organic acids, and antibiotics are such industrialized metabolic substances.

The simplest concept to understand is to separate water and ethanol by using the difference of boiling point. When yeast is cultured in a bioreactor and solids such as yeast and impure particles are separated from the liquid medium that ethanol is fermented, only water, ethanol, and small amount of by-product are left. The most general way to collect ethanol from such mixture is called distillation which uses different boiling points. By using distillation method, about 95% of ethanol concentration can be collected. When starch is used as the substrate, the concentrated alcohol is used as the drinking alcohol by dilution with water. With the use of the cellulose material, these are mostly used as transporting fuel for the automobile (Figure 8-4). All the water needs to be removed if the desired product is anhydrous ethanol. In this case, it is hard to get rid of the water by distillation method but it is more effective to use ultrafiltration that uses the membrane or adsorption that uses porous adsorbent.

Membranes are generally used to separate and concentrate a certain size of substance in the aqueous solution. Therefore, they can also be used to carry out continuous biological reactions with a cell or an enzyme immobilization and membrane attachment to the bioreactor. The membrane itself can also be modified to work as a bioreactor itself. The basic concept of the membrane uses separation-concentration by difference in water pressure. Membranes that are used for kidney dialysis separate materials by using concentration difference. One thing to consider when separating and concentrating a substance using a membrane is to understand how much flux per unit area can be introduced to separate and concentrate. It is also necessary to find a solution to physically or chemically remove the small materials that causes fouling of the pores in the membrane surface by completely or partially clogging the membrane pores. Membranes should be made of materials that can be sterilized because they can easily be contaminated by microorganisms and should last longer to be successfully used economically. Membrane materials are normally made of a variety of polymeric materials. The pores of the membrane can be easily clogged because the surface of the membrane is hydrophobic and the proteins that have hydrophobic characteristics can easily bond together, therefore, the property of the membrane surface could be changed to hydrophilic. The average pore size of the membrane is about 0.001 – 0.02㎛ which is much smaller, more uniform size, and higher operating pressure than the normal filtration systems with average pore sizes of 0.02 – 10㎛. Another distinct feature is that the membranes are operated with the cross-flow filtration system. The popular types of membranes used are plate and frame membrane, hollow fiber membrane, and spiral wound membrane. Plate and frame membrane is generally used in both laboratories and industries, hollow fiber membrane is popular in wastewater treatment, and spiral

Figure 8-5. Different types of membrane

Hollow-fiber membrane

Spiral-wound membrane

Flat sheet membrane

wound membrane is mainly used in desalination of sea water. The reason why these membranes can be used in wastewater treatment or sea water desalination is because the mechanism is rather simple and easy for scale-up (Figure 8-5).

Adsorption is a method used for concentration and purification of antibiotic or small molecular substances, and a certain amount of substance is adsorbed on a solid substance (adsorbent). Adsorption has two different types – physical adsorption and chemical adsorption. Physical adsorption is caused by weak mutual attraction between molecules and chemical adsorption occurs from chemical bonding between molecules. Most of the separation processes occur by physical adsorption. Normally, materials that are porous with large surface area are used for adsorption and they are used to separate gaseous or liquid mixtures, to purify biological products, or to remove impurities from the products. In the environmental fields, adsorption method is widely used for air purification or wastewater processes. An important scientist to remember when it comes to adsorption process is an American Physicist, Irving Langmuir (1881-1957). Langmuir proposed the concept of monolayer adsorption (adsorption isotherm of Langmuir) and was awarded the Nobel Prize in Chemistry in 1932. An adsorbent used in a small sized Korean seaweed snack is made of small beads. These beads have large porous surface area that effectively adsorb the moisture to keep the seaweed snack dry and crisp. In general, inorganic substances such as silica, alumina, and activated carbon or polymeric organic substances such as dextran and polystyrene or mixed materials with inorganic and organic substances are used as adsorbents (Figure 8-6).

As mentioned above, a desired material can be effectively separated when the physical and chemical properties of substances are used properly.

Let's hypothesize that an enzyme produced by a cell in a bioreactor is secreted into a liquid medium. Since most of the enzymes are soluble in

Figure 8-6. Materials for adsorption

Organic materials	Composite	Inorganic materials
Dextran, Polystyrene etc	Organic material + Inorganic material	Silica, Alumina, Activated carbon etc.

water, enzymes dissolved in the liquid medium can be separated as follows. Using the similar idea, adding a small amount of salt such as ammonium sulfate [$(NH_4)_2SO_4$] successively into the remained liquid medium after isolating the cell first, the enzyme is precipitated to the bottom due to the unstable characteristic of protein. The precipitated proteins are not 100% pure enzymes and still contains other enzymes and proteins, thus, precipitated proteins including enzymes are partially purified and are commercialized by being produced as a liquid or dried powder depending on its use. To further increase the purity of the enzyme, much higher enzyme activity can be obtained by using a suitable ultrafiltration membrane kit with consideration of the desired enzyme's molecular weight. Other than that, the enzyme can be extracted by using a solvent such as acetone, methanol, or ethanol. Extraction is a method that uses a specific substance as a solvent when the material is either solid or liquid. When the sample is solid, it is also called solid-liquid extraction or leaching. When the sample is liquid, it is called liquid-liquid extraction. Sometimes two immiscible solvents can be used based on the concept of the difference in solubility of those two solvents. Among the biotechnological products, methanol is most widely used as a solvent for medicine extraction because of its low cost and excellent extraction efficiency. Lately, extraction by supercritical fluid has been fairly popular in the food and pharmaceutical industries. The enzyme can be effectively extracted based upon the electrostatic attraction between the enzymes when appropriate organic solvents are used in each system. One thing to be careful in here is that the enzyme protein structure stays stable only when the operating temperature is maintained in between 0-5℃. If the temperature exceeds over 10℃, the tertiary structure of the enzyme protein is completely disintegrated. Another way to extract enzymes is by using polymeric materials. The most well-known method is called aqueous two-phase extraction or also called partitioning extraction. This method enables complete extraction while preserving the original shape of the protein, the enzyme, the cell and its components, and the virus. Not only that, this process can also remove the nucleic acids, polysaccharides and unnecessary cell debris that are produced after cell disruption during the separation of intracellular components. In general, two types of water-soluble polymers that cannot be mixed together are used or one

Figure 8-7. Aqueous two-phase extraction

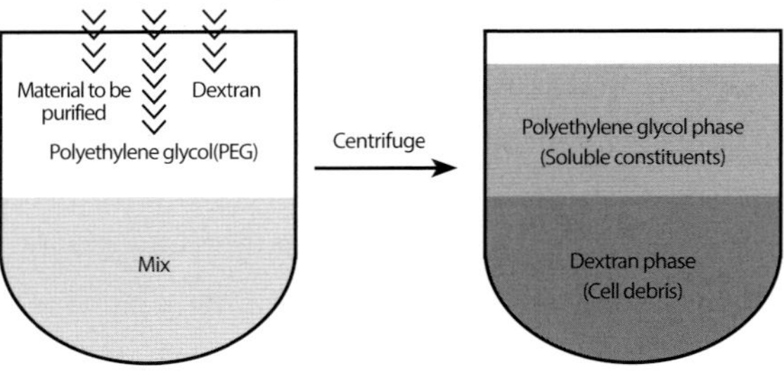

kind of polymer and a salt solution with high ionic strength that are immiscible could be used together. During this process, two different layers may be formed – solid particles in one layer and the soluble components as the other layer. A good example would be the use of polyethyleneglycol (PEG) and dextran after cell disruption. When each material is used in high concentration, the top layer made of PEG separates the soluble materials without mixing, and the bottom layer made of dextran can separate the cell debris. When both PEG and phosphorus are added to the cell culture solution, the enzymes are separated in the upper PEG layer and the cells are separated in the bottom phosphorus layer. After this, the enzymes to be separated are concentrated by the ultrafiltration membrane, and PEG can be recovered and employed for the next process (Figure 8-7).

If the enzyme that needs to be extracted is a pharmaceutical enzyme for a diagnostic or therapeutic purpose, the degree of purity must be much higher. To get the higher degree of enzyme purity during extraction, the structure and property of the enzyme should be thoroughly understood. During this process, good application of the liquid chromatography with strong foundation on various principles is needed. Much like the chemical separation process, filtration, centrifugation and extraction are frequently used during the initial stage of separation. Chromatography is normally used at the final stage for purification once all the material to be separated are concentrated and the impurities are mostly removed. If these steps are underestimated, the whole stages can become very costly due to the problems on the expensive columns. The classification according to the basic principle of the liquid chromatography could be an ion exchange chromatography that uses charge differences, hydrophobic interaction

Figure 8-8. Chromatography

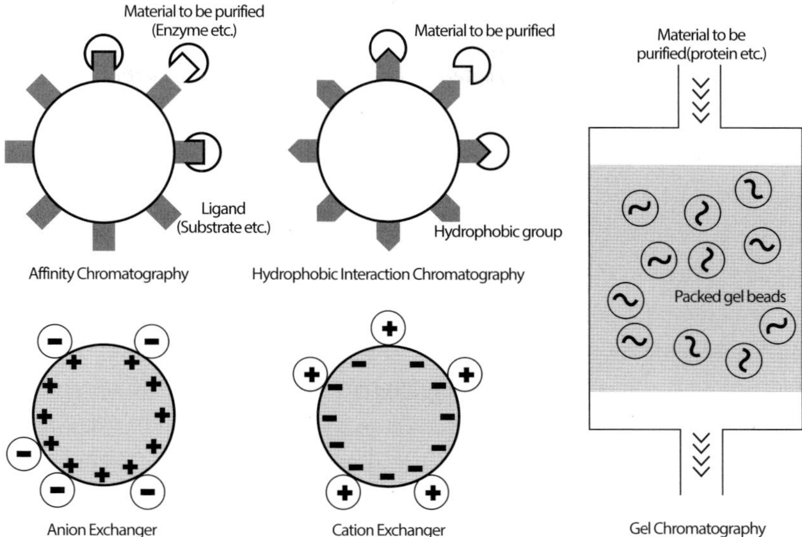

chromatography(HIC) that uses the hydrophobicity of material surface, affinity chromatography using the bindings of functional groups of substances, and gel chromatography that uses the size and shape of the material (Figure 8-8).

Two types of ion exchange chromatography exist – Anion exchangers and cation exchangers. Anion exchangers are attached with the beads that have the positively charged property on the surface which makes them capable of adsorbing negatively charged materials. Cation exchangers are bound with the negatively charged beads that attract and adsorb the positively charged materials. Once the materials to be separated are adsorbed, they can be collected by using a buffer solution containing sodium. This method is much cheaper and easier to perform than other chromatographic methods and is also easy to scale up more extensively. When proteins are being separated, the anion exchange resin is mainly used because most of the proteins carry the anionic property. The carrier used as an exchange resin is generally made of materials that are solid-hard, inert, and porous.

Hydrophobic interaction chromatography (HIC) separates hydrophobic materials with the beads that have the property of hydrophobicity on the surface so that the hydrophobic materials covalently bind with the hydrophobic beads. Since some amino acids that carry the hydrophobic characteristics are located on the surface of the protein, the protein can be separated due to this interaction. HIC is generally used in the fields such as pharmaceutical industry or wastewater treatment.

Affinity chromatography collects the desired materials by immobilizing various ligands onto the surface of the beads so that the materials to be separated can bind with the immobilized ligands. For example, when a substrate is attached as the ligand on the surface of the bead, only the desired enzyme can be collected due to the interaction made between the substrate and that specific enzyme. These types of reactions can be used not just from the substrate and the enzyme but also from the interaction between receptors and hormones or antigens and antibody responses. Therefore, affinity chromatography is a very selective method compared to others. After the reaction is performed at the column, unbound materials can be washed out with an appropriate buffer solution, and the bound materials can be collected with a different type of buffer solution that is specifically used for collection.

Gel chromatography has multiple different names such as gel filtration chromatography, size exclusion chromatography or molecular sieving. Gel chromatography uses porous gel beads that are manufactured by treating polymeric materials with a crosslinking agent. The beads contain a small maze or labyrinths inside that make small-sized objects to take a long time to passing through the beads. Larger-sized materials that cannot pass through the bead's inner maze rather pass through the column under a short period of time. Differently sized materials can easily be separated by using this method. Since the carrier used in this method is a lot more expensive and can elaborately separate very small objects such as salt with low molecular weight, this method is generally used in the very final stage. It usually takes about three to five final stages of chromatographic methods to separate antibiotics.

Separation of accumulated products inside a cell occurs mostly when the product is the recombinant proteins. The cell must be separated first before the breakage because the product is contained within the cell. Which cell disruption method to use to break the cell wall and membrane solely

Figure 8-9. Ultrasonic cell disruptor

depends on where most of the product are located inside the cell. There are multiple types of devices used for cell lysis. One method crushes the cell like how a pharmacist grinds the medicine with a mortar and pestle. Another method uses ultrasound like an optician cleans dirty materials on the surface of the glasses. Using these similar yet different methods, the cell can be completely crushed or broken down in a mixer-like equipment. There is also a method that pressurizes and breaks the cell while the cell goes through a narrow and long passage. Some cells can also be broken by being mixed with multiple small glass beads so that the cells can be broken when the glass beads are hitting each other. Depending on the characteristic of the product, an appropriate cell disruption method is chosen to minimize the loss of the desired product. After the cell disruption, the small fractured cell residues must be completely removed. Then, the subsequent purification steps should be efficiently designed according to the characteristics of the specific protein as in the case of protein purification (Figure 8-9).

Even if a recombinant protein is obtained, most of the proteins in this state do not have an appropriate activity. These proteins must go through protein refolding in order to regain the active structure, and subsequent separation with membrane or chromatography during this process is necessary for further purification. Therefore, these types of products are high value-added products and the cost of the product itself is a lot more expensive since the cost is higher for separation and purification than the other two types mentioned in previous cases.

Other than that, electrophoresis is an important technique used in laboratories to analyze, separate, and purify biopolymers such as DNA, RNA, and proteins. All biopolymers have their own charge depending on the size and shape, thus, each biopolymer moves in different rates when placed into a constant electric field. In the earlier years, a sugar solution was

Figure 8-10. Analysis of protein molecular weight with SDS-PAGE

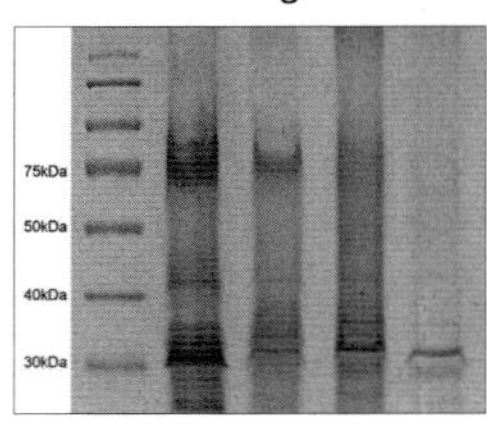

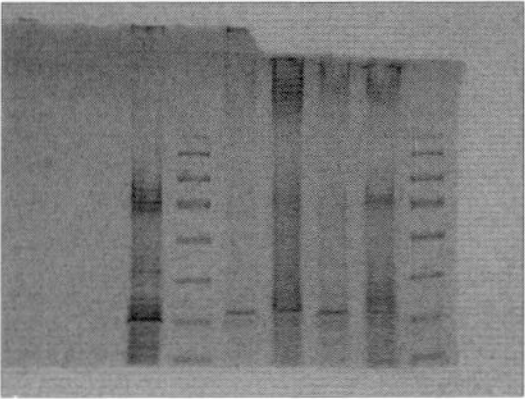

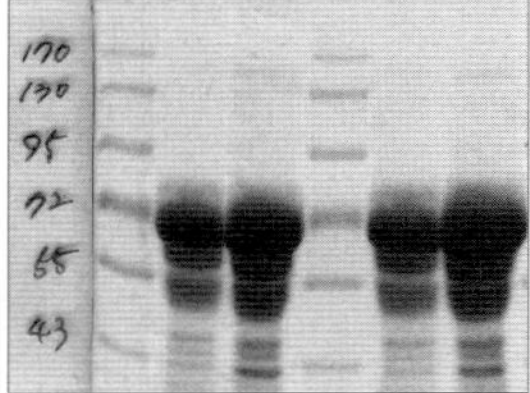

used to support for separation but nowadays mostly acrylamide or agarose gels are used. In general, sodium dodecyl sulphate (SDS) electrophoresis and isoelectrofocusing (IEF) are mainly used to accurately determine the molecular weight and the isoelectric point of the material. In here, the isoelectric point pl is the pH value when the net charge is zero, and the IEF enables the separation of different materials according to the difference of the isoelectic point. This method is very simple, cost-effective, and several samples can be analyzed at once. Electrophoresis also has a high sensitivity with a clear band-shaped result, and can be applied with enzymatic or immunological methods. Many of the times, electrophoresis using the polyacrylamide gel is performed. This is called polyacrylamide gel electrophoresis (PAGE). When the separated materials are visible as band-like results using PAGE as a support, Coomassie blue (Brilliant) or silver are normally used as the dye for confirmation. In SDS-PAGE, SDS is an anionic detergent that separates proteins into protein subunits so that the proteins are visible as a band-like shape. When using the acrylamide gel, the protein itself comes out as a band but using the SDS-PAGE allows scientists to obtain more information by enabling the multiple broken-down protein subunits to appear as bands (Figure 8-10). When both SDS-PAGE and IEF methods are conducted at the same time, two-dimensional gel electrophoresis is performed, which brings out much better result than when these methods are used separately.

The basic principle of DNA and RNA electrophoresis is same as the protein electrophoresis but DNA/RNA electrophoresis do not use SDS. The analysis of the double strand and the single strand is slightly different, but the composition of the polyacrylamide gel and the presence or absence of urea and formamide are different when the gel is polymerized.

Figure 8-11. Agarose gel electrophoresis for the separation of DNA fragment

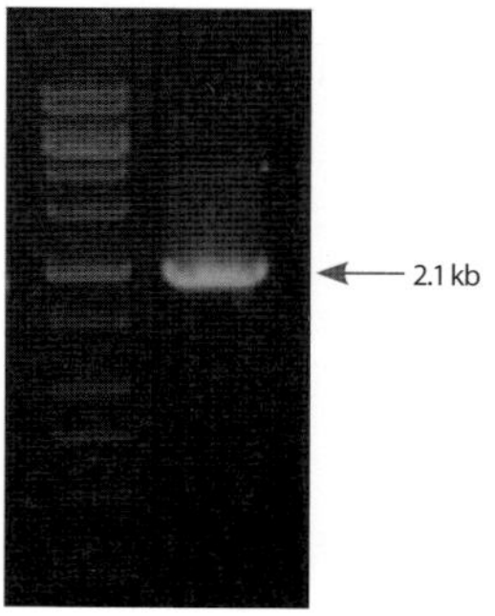

When identifying the nucleotide sequence, ethidium bromide (EtBr) is used to stain the nucleic acid (Figure 8-11).

The final stage of the purification step is called "polishing". At the polishing stage, crystallization and drying are the crucial steps to go through. Crystallization is used as the intermediate or the final step to separate and purify the highly pure products in food or pharmaceutical industries. For the efficient crystallization, most optimal condition must be established for the material. Especially, substances that are highly sensitive to heat must be performed under low temperature. Generally, properties of the product depend on the generation, growth, and aggregation of the crystal nuclei. However, the property of the crystal particle could be different depending on the distribution of the crystal size, structure and form, purity, or yield of the crystal. Therefore, the final product can have various crystal structures such as salt and sugar depending on the complex circumstances of crystallization. In the pharmaceutical industry, the cases of variations in drug efficiency are often observed according to the different crystal structures of drugs. Crystallization is normally performed with a batch or continuous operation in a crystallizer. This was applied from the principles of stirred-

Figure 8-12. Crystallization equipment

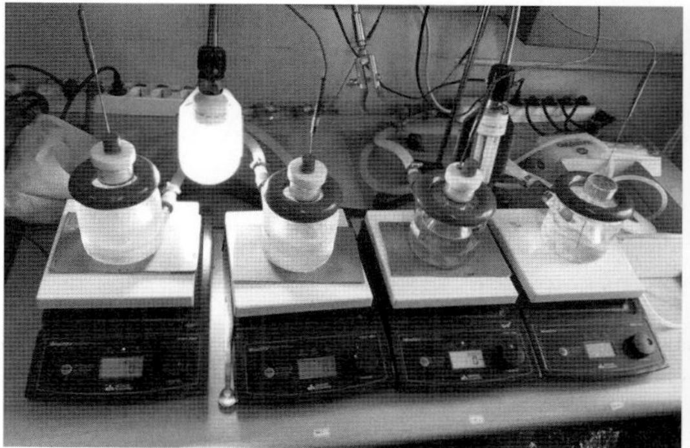

Figure 8-13. Drying equipments

Hot air dryer Freeze dryer

tank reactor or mixed systems of stirred-tank reactor and a fluidized-bed reactor (Figure 8-12).

Drying is used as the final stage in the manufacturing processes of the biotechnological or microbiological product in the biotechnology field. The temperature-sensitive products are normally dried under low temperature for a long time, and the temperature-resistant products are dried under high temperature for a short period. Drying the product reduces the transportation cost, easier to handle and pack, and easier for long-term storage and distribution. There are multiple types of dryers – vacuum-tray dryer, freeze dryer, spray dryer, rotary drum dryer, pneumatic conveyor dryer, and vacuum rotary dryer (Figure 8-13).

A vacuum-tray dryer is mainly used in the pharmaceutical industry and has the advantage to minimize the loss of drug efficacy and heat destruction when drying small amount of expensive products. A freeze dryer is widely used in the food industry by freezing and sublimate water from microorganisms, enzymes, and antibiotic suspensions under the vacuum condition. A spray dryer is normally used to dry the products that are sensitive to heat. It applies the way a fluidized-bed reactor works. The product suspension is sprayed through the nozzle unit at the top of the dryer and meets the hot gas that is supplied from underneath the dryer. As the product suspension and the hot gas meets, liquid evaporates and the dried product falls onto the very bottom of the dryer in a matter of seconds. Various sizes of the spray dryers exist and are chosen depending on the types of biotechnological products. When a rotary drum dryer is used, a product suspension is poured in between the two drums that are heated by hot steam. Then, the water evaporates as the two drums start to rotate inwards and the product adheres to the surface of the drums. The product falls to the bottom of the rotary drum dryer by the blades located on the outside of the drum and can be collected through the bottom. Since this dryer uses high heat, products that are heat-sensitive are not appropriate to be used with the rotary drum dryer. A pneumatic conveyor dryer uses hot air and dries heat-sensitive or easily oxidizing products within couple of seconds. However, big-sized or porous products are not suitable for the pneumatic conveyor dryer since this type of dryer forces the product to go through a long passage under the short period. A vacuum rotary dryer looks like the two cone-shaped containers glued together and has a jacket on the surface so that the water or fluid can heat up the container. The product is dried as the container rotates, but this dryer is also not very suitable for big-sized products.

As various types of dryers are used to dry the products, the percentage of moisture to be left on the product can be manipulated depending on the purpose of the product. Normally, the moisture is kept within 5% to increase stability and to prolong the distribution period. If the water content is high, the product is more likely to be damaged due to increased possibility of microbial contamination. The product that contains high water content can be sold immediately while refrigerated.

Figure 8-14. Examples of formulation

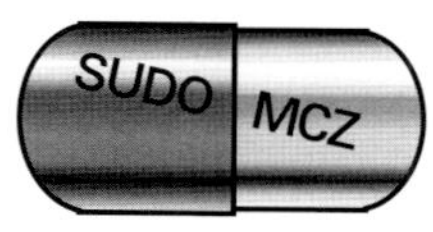

Mycosol capsule
Antifungal agent

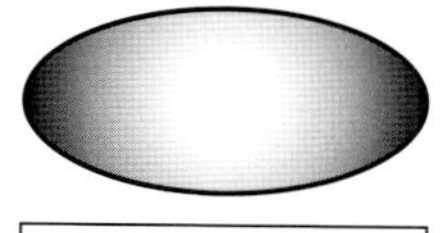

Choline alfoscerate capsule
Muscarinic receptor agonist

As the very final step, products go through a stage called formulation. The process of formulation maximizes the effectiveness of the active ingredient of the product. Though the reason to go through the formulation process differs by the purpose of the product, formulation normally allows the product to stay stable for longer distribution period to make the product to be used in a desired purpose such as treating diseases or eradicating parasitic pests on the plants (Figure 8-14). Let's look into some examples.

Figure 8-15.
Lactic acid bacteria cosmetics

Let's use food as an example. Yogurt sold on the markets can be called a formulation of lactobacillus (types of lactic acid bacteria). Lactobacillus resides inside the human or the animal's intestines and produces lactic acid by using saccharides to prevent infection and improves immunity. Therefore, it is used as a formulation agent in humans and used as feeding additives for animals. Lactobacillus is being popularly used lately in cosmetics, too (Figure 8-15).

Lactobacillus preparations are mostly sold in a liquid form, but powder products are also increasing for long-term preservation. For a commercial product, lactic acid bacteria are cultured in a large scale of bioreactor. As mentioned above, powder types of lactic acid bacteria can be obtained by freeze-drying or spray-drying at the end after going through several steps such as filtration and centrifugation. When lactic acid bacteria in the liquid or powder form is consumed, most of them do not survive to the large intestines due to the gastric acid and enzymes that exist in the stomach or small intestines. To keep the efficacy of the lactic acid bacteria maximum inside the body, it is formulated by coating with gelatin, carboxymethylcellulose (CMC), and trehalose or by microencapsulation with alginate and chitosan etc.. When the lactic acid bacteria are formulated with these materials, most of the bacteria stay alive and make its way to the large intestines and prevents infection and improves immunity. This is an application of the immobilization technique that is mentioned above.

Secondly, let's look into the medicine products. Assume that the pure penicillin is separated through appropriate separation methods after being produced in a large scale of bioreactor. The pure penicillin must go through formulation in order for it to be used to a patient. When the oral administration is used for medicine to treat certain diseases, the structure of medicine is broken down by gastric acid which prevents the medicine from showing its full efficacy. That is why formulation is necessary in medicinal products. Therefore, the encapsulation method that is part of the immobilization method mentioned above, is a good method to use in this case. When dried and powdered penicillin is encapsulated with a special polymer material, the encapsulated penicillin can withstand the gastric acid with very low pH, and safely arrives to the desired destination and starts to be slowly diffused inside of the small or large intestines as the polymer material is broken down by the higher pH of the intestines.

The third example is the spores of the fungus that are used as biological pesticides. Assume that we are using a formulated environmentally friendly fungal spores as biological pesticides instead of using chemical pesticides to kill the larvae of parasitic insects on some plants. If the larvae are infected with the fungal spores, they cannot stay alive. Same with other products, the fungal spores cannot be used as pesticides right after being produced in a large scale from a bioreactor. The spores must go through formulation stage and be prepared as a form of suspension. As the suspension is sprayed to the larvae on plants, the spores can infect the larvae and kill most of them. The important part to remember is that the dispersant that allows fungal spores to disperse thoroughly, the minimal nutrition such as saccharides, and the UV protector so that the spores can be protected from the strong sunlight must be added at proper concentration during the formulation process. Immobilization technique is applied to enable these functions.

Lastly, the proteolytic enzymes used as the enzyme detergent cannot be commercialized without going through the formulation process. Enzymes are first produced in a bioreactor, precipitated by adding salt into the liquid medium or are separated with the membrane, and then one of the immobilization technique is used for formulation. This allows the enzymes to crosslink together naturally or by adding the crosslinking agents. As the enzymes are formed as very small beads, they can be used as the detergent additive with more addition of surfactant, bleach, fragrance, and multiple other ingredients.

As shown in this chapter, the mass-produced products from the bioreactor must be well separated, purified, and formulated through economically designed downstream processing in order to make a highly valuable product.

Chapter 9

Bioresources and Medical Biotechnology

Bioresources are the core material of the bioindusty and are widely used in the fields of medicine, food, energy, and environment. As the ‹Nagoya Protocol› was entered into force on October 12th, 2014, the importance of bioresources started to become more conspicuous. As the ‹Nagoya Protocol› was adopted, it became necessary to obtain prior informed consent from the relevant agency of the resource-holding country and conclude the mutual agreement under certain condition after meetings in order to use the genetic resources and related knowledge of other countries. As utilizing a different country's bioresources is thought to become more challenging, it is important to continuously try to keep the proper relationship with other countries to secure more bioresources. Not only that, effort to finding new biological resources while preserving the domestic bioresources is much needed at this point. Chapter 2 mentioned briefly about the basic principle and application of bioresources, and the other chapters briefly talked about different areas that use bioresources. Since the bioindustry field that is based on biotechnology with the use of bioresources is very broad, we will be mostly focusing on the biomedical industry in this chapter.

As mentioned above, bioresources include microorganisms, plants, animals, and their genes, proteins, functional materials and information produced from these organisms. Another important type of bioresource is human-derived material. Microorganisms such as bacteria, archaeans, fungi, virus, mushrooms, or Cordyceps are widely used as bioresources of industrial application. Plant-based bioresources include seeds, plant cells and tissues, and genetically modified plants. In animals, laboratory animals, animal tissues, cancer cells and stem cells, cell lines such as hybridomas, fertilized eggs, and genetically modified animals are used as bioresources. As of human-derived material bioresources, such as human tissues, cells, cell lines, and human-derived materials and information can be used as long as they do not violate bioethics.

By using such various bioresources and biotechnology techniques, the field of biotechnology has been immensely growing over the years. As Chapter 1 classified biotechnology as medicine, industry, agriculture and marine biotechnology, bioindustry can also be organized as industries of medical biotechnology, industrial biotechnology, agricultural biotechnology, and marine biotechnology. Plus, the fusion biotechnology industries that is formed by fusion of two or more industries are starting to emerge. In here, we will look into the bioindustry fields with more detailed manner.

In 2008, South Korea organized the areas of bioindustry into eight for clear definition. The eight areas of bioindustries are: biophamaceutical industry, biochemical industry, biofood industry, bioenvironmental industry, bioelectronics industry, bioprocess and equipment industry, bioenergy/bioresource industry, and bioinformation service and research/development industry.

According to the data from Korea Biotechnology Association (2013), the status of human resources of each bioindustry area shows that about 45% of human resources are under biophamaceutical, 23.5% in biofood, 14% in biochemical, 4.7% in bioassay, 3.4% in bioenvironment, 3.3% in both bioenergy/resources and bioprocess and equipment, and 2.7% in bioelectronics.

The followings are the details of each bioindustry category.

- Biophamaceutical: Antibiotics, anticancer agents, antiviral agents, biosimilars, vaccines, gene therapy, cell therapy, in vitro diagnostic devices, etc.
- Biochemical: biodegradable polymers, industrial enzymes, organic acids, amino acids, microbial leaching, pigment, fragrance, etc.
- Biofood: fermented food, functional food, substitute sweeteners, food additives, genetically modified crops, antioxidants and functional food derived from microalgae, etc.

- Bioenvironmental: bioremediation, microbial formation, environmental monitoring, biological desulfurization or deodorization, etc.
- Bioelectronics: bioelectronic nose, bioelectonic tongue, etc.
- Bioprocess and equipment: fermentation process, bioseparation process, animal and plant cell culture, biotransformation, bioreactor, etc.
- Bioenergy or bioresources: bioenergy, immobilization of carbon dioxide, etc.
- Bioassay: biosensor, biochip, etc.

As shown above, there are various types of products produced by using bioresources, and the eight areas of core technology used are mentionted above. This chapter will mainly focus on the biopharmaceutical industry which is the largest biotechnology market of all in the world, and Chapter 10 will focus on the production of various biotechnological products using biomass as a future of biochemical industry with much potential that will gradually replace the current chemical industry.

The biopharmaceutical field is one of the largest sales areas in the bioindustry. The biopharmaceutical field can be divided into two parts: chemically synthesized pharmaceuticals with low molecular weight such as antibiotics, anticancer agents, antiviral agents and biopharmaceuticals such as proteins, antibodies, vaccines, gene therapy or cell therapy.

Antibiotics are substances produced by microorganisms that can inhibit the growth of other microorganisms or completely kill the other microorganisms with the minimum concentration. Up to now, most of the industrialized antibiotics have been produced by about 60% of actinomycetes and 20% of mold. The penicillin mentioned in the previous chapter is the first industrialized antibiotics which acts as an inhibitor for other pathogens to fail the cell wall synthesis. Antibiotics are generally associated with mechanisms of inhibiting the synthesis of cell wall, protein, or nucleic acid or changing the cell membrane permeability or inhibiting the cell metabolism. This type of mechanism will depend on the structure of the antibiotic, and different types of antibiotics can be distinguished based on the structure. The narrow spectrum antibiotics only affect either gram-positive or gram-negative bacteria at once, and the broad spectrum antibiotics can influence both types of bacteria (Figure 9-1).

Figure 9-1. Classification and mechanism of antibiotics

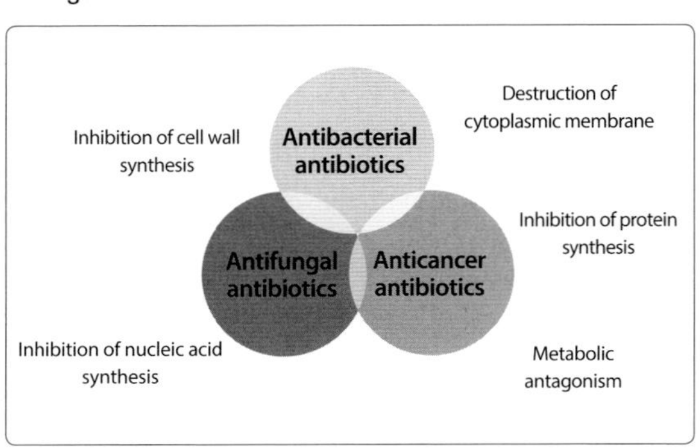

Currently, the rise and increase of the antibiotic-resistant bacteria is becoming a major issue all around the world. The measure of bacterial resistant to antibiotics is calculated by the number of bacteria surviving out of a hundred bacteria after the introduction of antibiotics and this is expressed as the antibiotic resistance percentage. The main reason why bacteria build antibiotic resistance is because of an amazing adaptability of microorganisms, mutation due to evolution, imprudent selling of antibiotics before separation of prescribing and dispensing drugs, antibiotic overdose by the doctor, and overuse of antibiotics into the animal feeds and plant disinfectants. The most well-known multidrug-resistant bacteria are methicillin-resistant *Staphylococcus aureus* (MRSA) which methicillin is an alternative treatment of penicillin, vancomycin-resistant *Staphylococcus aureus* (VRSA) which vancomycin can treat infection by MRSA, vancomycin-resistant *Enterococcus faecium* or *E. faecalis* (VRE) which is intestinal bacteria, imipenem-resistant *Pseudomonas aeruginosa* which is resistant to imipenem that is widely used as a broad-spectrum antibiotic, and imipenem-resistant *Acinetobacter baumannii* which is a 'superbacteria or superbug' that has strong resistance towards imipenem. Most of these strains cause skin diseases, organ infections on lungs or kidneys, urinary tract infections and at worst, a severe sepsis in patients with weak immune system.

In March of 2015, the US government with the former president Barack Obama announced the National Action Plan to Combat Antibiotic-resistant Bacteria. The Center for Disease Control (CDC) released that about 2 million patients arise and about 23,000 patients are deceased every year due to antibiotic-resistant bacteria. The CDC reported that these following antibiotic-resistant bacteria have the highest risk of all – *Clostridium difficile* that is resistant to fluoroquinolones; Enterobacteriaceae which has resistance to most antibiotics including carbapenem; *Neisseria gonorrhoeae* that is resistant to cephalosporin. The National Action Plan to Combat Antibiotic-resistant Bacteria project provides a blueprint of possible solutions to the nations with antibiotic-resistance problems, support for an international activity to prevent infections from antibiotic-resistant bacteria and the information on efficacy maintenance of conventional and new antibiotics, new method for infection diagnosis, and new drug development including vaccines.

The Bangkok Post posted a column titled the "Drug resistance grows menacingly" on December 21, 2015. The column includes the information that 1 out of 5 children die due to antibiotic-resistant bacterial infection in Southeast Asia, and it is becoming worse every year. About half-million human beings die around the world every year due to the infection from antibiotic-resistant bacteria.

South Korea has been trying to decrease the bacterial antibiotic resistance by establishing a system to standardize the accurate antibiotic resistance information at the Center for Disease Control. However, the government still has to build the stronger systematic policy and endeavor into the new antibiotic development since the number of bacteria that are resistant to antibiotics are increasing.

The medical term for cancer is tumor, and the study of tumor is called oncology. The vocabulary 'cancer' arose from a Latin word meaning a crab. This is because crabs do not let go of anything once they pinch with their claws. Cancer can be classified into two different types: a benign tumor and a malignant tumor. A benign tumor grows big but in a slow rate, and does not spread out. A malignant tumor grows fast, grows big as it spreads out to different tissues in the body. Tumors normally arise due to mutation of the genes of a single cell, and the mutation causes the cell to change its size and structure. This type of cell does not have a constant number of chromosomes, goes through frequent cell division, and all these cannot be controlled. They also lose the shape in between cells. Tumors can arise due to the accumulation of DNA damage and errors of a cell. Of course, genetic factors play a role in having a tumor, but environmental factors such as ultraviolet rays, radiation, toxic chemicals, or virus can take a part also.

According to the 2013 statistics by the National Cancer Information Center of South Korea, the highest cases were thyroid cancer, stomach cancer, and colon cancer in order. Males mostly had stomach cancer, colon cancer, and lung cancer in order and females mostly had thyroid cancer, breast cancer, and colon cancer in order. The 2014 statistics show that the cancer types with the highest death rate for men were lung cancer, liver cancer, and stomach cancer in order. For women, lung cancer, colon cancer, and stomach cancer had the highest death rate in order.

Normally, the cancer treatments are performed by surgery, radiation therapy, chemotherapy, and biopharmaceuticals. A surgery is the oldest method that has been used by doctors with addition of other treatments. The radiation therapy kills the tumor cells by destroying the division and growth of tumor cells with high-energy radiation. However, this method can also have a negative effect to normal cells. The strength and dose of radiation can be determined by the size, type, range of area, differentiation degree, and reactivity of the tumor cell. The chemotherapy uses anticancer drugs with different mechanisms. For instance, one of the anticancer drugs can inhibit DNA replication by binding itself to the DNA of tumor cell, but it has side effects such as hair loss, vomiting, diarrhea, physical pain, or bacterial infection because this mechanism also works in normal healthy cells. Treatments using biopharmaceuticals kill tumor cells by activating white blood cells inside of the body by using interferons or interleukin-2. A granulocyte-colony stimulating factor (G-CSF) can be used to enhance the activity of immune system that had lowered due to chemotherapy. As more cancer metastasis mechanisms continue to be identified, many researchers are developing the enzyme inhibitors to inhibit cancer metastasis.

More than 80% of the anticancer drugs that are being used currently are produced from actinomycetes. Although the physiologically active substances derived from actinomycetes continues to be screened, only the same substance gets separated with the same conventional method. That is why the molecular biological methods such as enzymes related to metabolic pathways and gene expression mechanism have been used with the traditional existing methods. Recently, the research on developing new anticancer drugs and antibiotics or increasing the productivity by inhibiting the entire metabolic pathway has been very active as the whole genome of the actinomycetes has been elucidated.

One of the most popularly used anticancer drug is called doxorubicin, which is a substance of anthracycline. This drug kills the tumor cells by inhibiting the growth and division of the tumor cells. It is also called as adriamycin, which is a cell cycle-selective anticancer drug that inhibits cell division. The alkylating agents such as cisplatin and carboplatin inhibit the cell division during the process of DNA replication and transcription, and are cell cycle non-selective anticancer drugs.

There are also plant-derived anticancer drugs. Substance such as taxol or paclitaxel derives from the bark of the tree called *Taxus brevifolia*. To extract taxol, an old yew tree must be cut off, which is not preferred. Currently, taxol is being produced by the plant cell culture and the semi-synthetic method which is produced by chemical synthesis after the precursor extraction from European yew tree. A good news is that taxol production through mass culture is possible as the researchers found out that fungus called *Taxomyces andreanae* could be isolated from the stem of the yew tree. Taxol is known to be effective in the treatment of ovarian, lung, and breast cancers (Figure 9-2).

Figure 9-2 Yew (Taxus sp.)

The targeted anticancer drugs were also developed based on the fact that cancer cells have different cell-cell signaling system and use different types of receptors compared to the healthy cells. It can be considered as a customized treatment for each patient since this treatment regards the genetic variation of cells or acquired environmental factors of each person. An epidermal growth factor receptor inhibitor called Iressa (gefitjnib) that minimizes the side effect towards normal cells and an angiogenesis inhibitor called Avastin (bevacizumab) are some of those targeted anticancer drugs. These types of anticancer drugs are considered as antibody drugs and newly developed products are continuously being introduced to the market. Currently, the sales of the targeted anticancer drugs have the largest market compared to other treatments, and the growth rate is immensely increasing. As the pharmaceutical industries of the United States, Europe, and Japan have been the major players of the targeted anticancer drug development, the market is considered to grow even larger in the future.

Figure 9-3. Causes of cancer

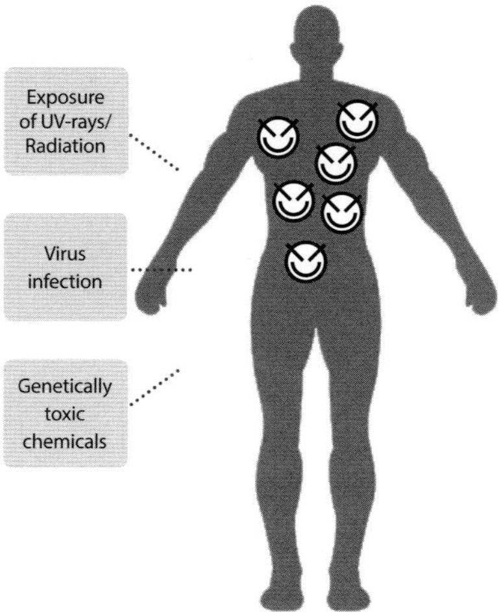

As the mortality rate of cancer is very high due to the metastatic character of the tumor cells, much research is currently being performed. Especially, developing new medicine based on the research identifying the factors that can predict tumor cell metastasis and the mechanism of metastasis itself has been very actively conducted (Figure 9-3).

In addition, anticancer drug resistance in patients is becoming a serious problem. The patient is thought to have anticancer drug resistance when the tumor cell still remains after a high enough dose of drug is administered. Though much information have been found on the mechanism of anticancer drug resistance, it is complicated due to the difference in each patient. Though many new anticancer drugs are being developed continuously, it is important to put steady effort in finding solutions for anticancer drug resistance.

For reference, the very first new drug developed in Korea was a stomach cancer drug by SK Chemicals in 1999. After developing the third-generation platinum complex anticancer drug called Sunpla, a total of 26 Korean domestic new drugs have been registered by December, 2015. Out of these 26 drugs, 5 of them are anticancer drugs, 4 antibiotics, 4 diabetic drugs, 3 arthritis drugs, 3 erectile dysfunction drugs, 2 antiulcer drugs, 1 hypertension drug, 1 malaria drug, 1 hepatitis B drug, and 1 *Pseudomonas aeruginosa* vaccine (Figure 9-4).

Figure 9-4.
New drugs developed in Korea

Product	Number of items
Hypertension drug	1
Anticancer drug	5
Hepatitis B drug	1
Malaria drug	1
Pseudomonas aeruginosa vaccine	1
Erectile dysfunction drug	3
Antiulcer drug	2
Diabetic drug	4
Antibiotics	4
Arthritis drug	3

Korea Drug Research Association(KDRA) - (1999-2015)

Virus-derived diseases such as acquired immune deficiency syndrome (AIDS), hepatitis, influenza, and herpes are continuously increasing all around the world. As diseases due to the mutation of viruses increase, the need for development and research on antiviral treatments is becoming more urgent. In addition, a solution to the resistance built against antiviral treatment is very much needed.

The antiviral treatment generally treats the diseases caused by viral infection and diminishes or destroys the action of the virus infecting the human body. It can also be classified as antibiotics in a broad sense. Currently, most of the viral infections are being treated with immunotherapy that uses vaccines. Zidobudine (ZDV), also called as azidothymidine (AZT), is used as a treatment for AIDS. ZDV interrupts DNA replication of the HIV virus by inhibiting the reverse transcriptase that is necessary in DNA synthesis of HIV. Tamiflu, also called as oseltamivir, is a treatment for influenza. Tamiflu is made from a substance called shikimic acid extracted from the native Chinese plant's fruit called *Illicium verum*, which is mostly known as a star anise, a Chinese spice. The shikimic acid goes under chemical synthesis for development of Tamiflu (Figure 9-5). Shikimic acid is mainly produced as an intermediate during the plant's metabolic pathway. Tamiflu was first developed by an American venture company named Gilead Sciences, and is being sold exclusively by the Roche company in Switzerland. Interferons (IFN) that help with immune system activity is used as a hepatitis treatment. This is a type of glycoprotein called cytokine that is made by the immune cells. There are also treatments called acyclovir and valacyclovir that are mainly used as herpes or skin disease treatment.

Figure 9-5. Star anise (*Illicium verum*) for Tamiflu

Middle East Respiratory Syndrome (MERS) outbreak in South Korea in May of 2015 was a big problem. MERS is caused by a RNA virus called corona virus infecting the respiratory and digestive system. Through this experience, we saw how important it is to continuously establish the prevention system and new treatments against various viral infections.

According to the news carried by the Korean Pharm Business Journal on July 5, 2012, the Global Business Intelligence (GBI) Research predicted that the market on generic antiviral drugs would grow up to 30% due to the patent term expiration in 2018. Out of all products, the treatment drug for HIV, a retrovirus that destroys human immune system and the cause of AIDS, was expected to have the highest market share.

The biosimilar that was briefly mentioned in Chapter 5, is considered to rapidly grow in the markets due to the expiration of the patent term of other various biopharmaceutical products that are in the current market. In addition, biobetter, a super biosimilar, has the concept of a me-too drug (or follows-on drug) by improving the efficacy and the side effect of the original biopharmaceuticals. Though, building security of its related research and the original technology is much needed. The biosimilar that is produced industrially, is mainly the biopharmaceuticals related with therapeutic proteins and antibodies (Figure 9-6). Therapeutic proteins are normally produced

Figure 9-6. Types of biosimilar

Protein biopharmaceuticals	**Antibody biopharmaceuticals**
Cytokines Hormones Therapeutic enzymes In vivo factors	Monoclonal antibody Receptor-antibody fusion proteins

through gene recombination and cell culture, and can be divided into the types of cytokines, hormones, therapeutic enzymes, and in vivo factors. Interleukin-2, G-CSF, and interferons are the types of cytokines; insulin, human growth hormone (hGH), and erythropoietin (EPO) go under hormones; therapeutic enzymes include tissue plasminogen activator (tPA) and Factor VIII; epidermal growth factor (EGF) and insulin-like growth factor (IGF) are included in in vivo factors. In here, cytokines are physiologically active substances mainly produced by white blood cells that regulate physiological functions or information transmission especially in the immune system. An erythropoietin or EPO is a glycoprotein hormone that is a precursor to erythrocytes and promotes the erythropoiesis (generation of red blood cells). It is used as a treatment for chronic anemia. A tissue plasminogen activator or tPA is a type of therapeutic enzymes that is generally used as a thrombolytic agent. Factor VIII is a blood coagulation agent that is used to treat hemophilia. An epidermal growth factor or EGF is an in vivo factor that stimulates the growth, proliferation and differentiation of cells and is mainly used as burn treatment or in functional cosmetic products. An insulin-like growth factor or IGF has a similar structure as insulin which is very important for growth in children and physiological maintenance in adults. With this characteristic, it is mainly used as a treatment for malnutrition.

Antibody drugs include monoclonal antibodies and receptor-antibody fusion proteins such as Rituxan, Humira, Remicade, or Enbrel. The main ingredient for Rituxan is called rituximab, which is mainly used as treatments for short-term chronic lymphocytic leukemia and autoimmune diseases. Humira (adalimumab) is used as treatments for chronic immune-mediated inflammatory diseases such as rheumatoid arthritis, ankylosing spondylitis, Crohn's disease, ulcerative colitis, and psoriasis. Remicade (infliximab), similar to Humira, is mainly used as treatments for Crohn's disease, ankylosing spondylitis, ulcerative colitis, and psoriasis. Enbrel is also used as a treatment for autoimmune disease such as rheumatoid arthritis. Rituxan uses a mechanism of inhibiting B cells, and Humira, Remicade and Enbrel use a mechanism of inhibiting a tumor necrosis factor (TNF).

Currently, many businesses around the world are establishing and approving the guideline to various products such as insulin, growth hormones, monoclonal antibodies, and G-CSF etc.. The critical part in technology is to establish the upstream, midstream, downstream technologies and CGMP strongly, and to strategically examine the clinical models and their action mechanisms. The leading biosimilar businesses in South Korea are Celltrion, Samsung Biologics, Samsung Bioepis, and LG Chem.

The concept of vaccine was first introduced by a British doctor named Edward Jenner (1749-1823) when he developed a preventive vaccine against small pox, which was a prevalent epidemic at that time. The purpose of the vaccine is to prevent diseases caused by pathogens such as bacteria and viruses in humans or animals by injecting the antigens into the body to form the antibodies that can fight against the pathogens. With production of antigens by decreasing the toxicity of pathogens by treating part of the pathogen or the toxin component itself, it can be used as a preventive medicine. To use the vaccines successfully as preventive medicine, safety, efficacy, stability, convenience, and cost must be carefully considered during the process of vaccine development.

Vaccines are considered as preventive medicine that can fight against only bacteria, only viruses, or both. The most widely used antiviral vaccines are against influenza, hepatitis, poliovirus, chickenpox, shingles, hemorrhagic fever, Japanese encephalitis, rubella, measles, mumps, pustules, yellow fever, and rotavirus. Antibacterial vaccines generally fight against diphtheria, tetanus, pertussis, typhoid, laptospira, cholera, pneumonia, Bacille de Calmette-Guerin (BCG), meningitis, tuberculin, and *Clostridium botulinum* toxin. *C. botulinum* is an anaerobic bacterium that produces toxins that constrict or paralyze an animal's muscles. These toxins are actually widely used in humans as cosmetic purposes after being separated and purified. The product called botox is used as a treatment to remove facial wrinkles by temporarily paralyzing facial muscles.

Vaccines can be classified as live attenuated vaccines (LAV), inactivated vaccines, toxoid vaccines, subunit vaccines, recombinant vaccines, vector vaccines, and DNA vaccines. A live attenuated vaccine is a vaccine that uses attenuated pathogen to induce the same immune response as an infection. An inactive vaccine uses pathogens that are inactivated by chemicals or heat. Most of the inactive vaccine is normally used with im-

Figure 9-7. Classification of vaccines

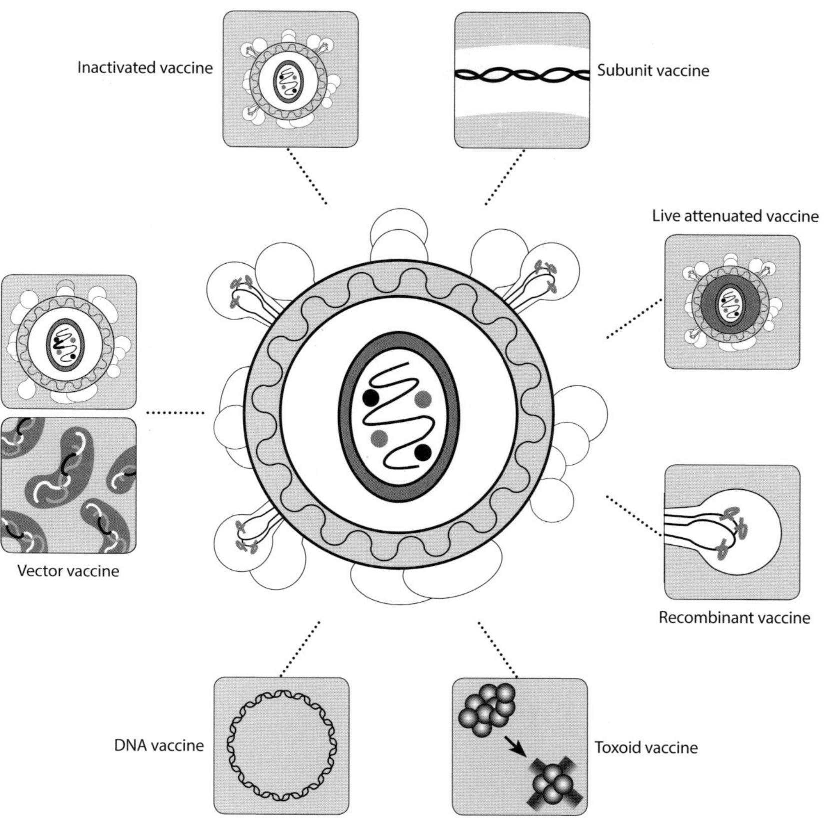

munostimulants as adjuvants due to its decreased immune effect. A toxoid vaccine induces antibodies by chemically or genetically inactivating toxins produced by pathogens. Diphtheria and tetanus vaccines are the examples of a toxoid vaccine. A subunit vaccine is used with an extract of antigens from a pathogen and is normally used with an immunostimulant as an adjuvant like an inactivated vaccine due to the reduced immune effect. A polysaccharide vaccine is also included in here. A recombinant vaccine isolates antigen genes from a pathogen, produces antigens by using gene recombination, and uses the antigens after separation and purification. Many of the subunit vaccines are also produced with gene recombination technology. A vector vaccine isolates the antigen genes from a pathogen, inserts into the virus and is used as a vector. A DNA vaccine injects the genetically engineered DNA of an antigen into a human body (Figure 9-7).

The Korean domestic companies that import or produce vaccines include the Green Cross, the Korea Vaccine, SK Chemicals, CJ Cheiljedang, LG Chem, Boryeong Biopharma, and Berna Biotech Korea.

Currently, the domestic research fields required for vaccine development in Korea include discovery of new antigens, development of immunity enhancer, basic research on immunology, development of vaccine administration pathway and delivery system, development of biological index for evaluation, and development of upstream, midstream, and downstream technology for research development and manufacture.

In October 8, 2015, the "Health Korea News" issued an article that Bill Gates, the founder of Microsoft, was investing towards a gene therapy start-up company. As a matter of fact, gene therapies and cell therapies that are considered as technologically advanced drugs, had some technological problems. Unlike other biopharmaceuticals, gene therapies and cell therapies inject active human genes or alive cells into the patients, which can cause more problems. However, as these technological problems have been solved by going through clinical trials, gene therapies and cell therapies have been entering the new era of commercialization.

Gene therapeutics are preventive medicine that treat the gene deficiency by injecting normal genes or cells with normal genes into the human body. A gene therapy can be classified into the somatic gene therapy and the germline gene therapy. The somatic gene therapy, which genes to be treated gets inserted into the somatic cells, is currently under extensive research. However, the germline gene therapy is strongly prohibited by law worldwide because of the risk of the inserted gene being transferred to the next generation.

The first gene therapy was performed by the National Institutes of Health (NIH) of America in 1989. The tumor infiltrating lymphocytes were collected from the tumor cell tissue of a patient with a stomach cancer. As the tumor infiltrating lymphocytes were cultured, inserted with the tumor necrosis factor (TNF) gene, and re-introduced to the patients, the tumor cells of the patients were partially reduced. Through this treatment case, the NIH confirmed the safety of the gene therapy.

One of the most prominent gene therapy is performed against adenosine deaminase deficiency (ADA). ADA is an inherited genetic disorder which the body cannot produce the enzyme called adenosine deaminase, and the body toxins start to build up due to the failure of toxin break-down. The toxins start to damage the immune-related cells and may cause severe combined immunodeficiency (SCID). For the patients, aseptic environment must be maintained completely to protect the infection from any pathogens. Overall, this disorder is also called as ADASCID. The gene therapy for ADA was the first successful gene therapy performed by the National Institutes of Health of America in 1990. The immune cells from the ADA patient were collected, and the normal ADA gene was injected into the patient by using a retrovirus. The patient introduced with this method was able to produce normal adenosine deaminase, and the immune system was also restored.

In 2012, a Dutch Company UniQure developed a gene therapy product called Glybera to treat lipoprotein lipase deficiency (LPLD) and was approved by the European Medicines Agency (EMA). LPLD is known to be a rare genetic disease of lipid blocking the blood vessels.

Other than that, the Sibiono Gene Tech Company in China commercialized a head and neck cancer treatment called Genedicine using the adenovirus vector in 2003 after the approval of the China Food and Drug Administration(CFDA). In 2005, the Shanghai Sunway Biotech Co. Ltd. also got approved by the CFDA with the stage 4 nasopharyngeal cancer gene therapy treatment called Oncorine. The Philippines and Russia also have the case of gene therapy approval. In Korea, the Kolon Life Sciences Inc., BioMed, Genexine, SillaJen, the Green Cross, and the Gene One Life Science Inc. are going under clinical trials of various gene therapies.

An important fact to consider during developing gene therapeutics is to clearly understand the characteristics of the cell that is to be used for gene therapy, and disease pathogenesis and its mechanism. Also, an appropriate gene to be used for gene therapy should be developed and a gene delivery method must be established. Researches related to this have been very active all around the world.

A cell therapy is a treatment that directly administers the cell therapeutics which includes living cells into the patient. The Ministry of Food and

Drug Safety(MFDS) defines the cell therapeutics as follows:

> A cell therapeutic is a medical product for the purpose of treatment, diagnosis, and prevention by changing the biological characteristics of the cell by using a method of multiplication and selection of living autologous, allogenic, or xenogeneic cells outside the body to restore the function of cells and tissues.

In here, an autologous cell means a self-originated cell, an allogenic cell means cells derived from other individuals, and a xenogeneic cell is an animal-derived cell. Most of the cell therapeutics are considered as custom-made medicines because they are re-injected to the patient after certain cells are collected from the patient, appropriate handling, and culturing.

Depending on the type of cells and degree of the differentiation, cell therapeutics for the cell therapy can be divided into the somatic cell therapeutics and the stem cell therapeutics. The somatic cell therapeutics generally include dermal cell therapeutics such as keratinocyte and fibroblasts, immune cell therapeutics such as dendritic cells, natural killer (NK) cells and lymphocytes, bone cell therapeutics, cartilage cell therapeutics, and adipocyte therapeutics. The keratinocyte therapeutics is mainly used as burn treatment, the fibroblast therapeutics in diabetic foot ulcer, wound care, and burn treatment. The immune cell therapeutics use dendritic cells, NK cells, and activated lymphocytes obtained from the patient's blood sample to increase the number of these cells and to strengthen the effect of the immune system. With this characteristic, immune cell therapeutics are widely used to fight against cancers and autoimmune diseases. In here, a dendritic cell is a type of an antigen-presenting cell that acts as an immune-response mediator.

In 2010, the biotechnology company called Dendreon developed and commercialized the world's first dendritic cell therapeutics for the prostate cancer. The NK cells are involved in the innate immunity and become matured in the liver and bone marrow. The activated lymphocytes are also called as T cells, and are mainly involved in antigen-specific adaptive immunity. The bone cell therapeutics promote the local bone formation in

Figure 9-8. Types of stem cells

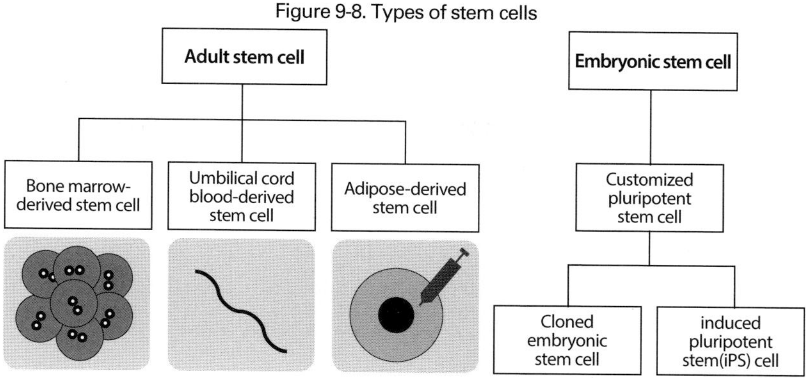

consequences of fractures, and the cartilage cell therapeutics are normally used when the knee cartilage is defective. After the global permission, more than ten products are sold in the Korean market such as diabetic foot ulcer treatments called Dermagraft and Apligraf by Organogenesis Inc., a burn treatment called Transcyte by ATS/Smith & Nephew, a chondrocyte-based cartilage cell therapeutic called Carticel by Genzyme Biosurgery, and Bioseed-C by BioTissue Technologies GmbH.

About ten cell therapeutic products are developed by the Korean domestic companies and approved for selling, and there are many more products that are still under the clinical trials. The products approved by MFDS are the cartilage cell therapeutics called Chondron and the bone cell therapeutics called RMS Ossron by Sewoncellontech Co., Ltd., the adipocyte therapeutics called Adipocell and Queencell by Anterogen Co, the dermal cell therapeutics called Holoderm and Kaloderm by Tego Science. The therapeutic product examples using the immune cells are dendritic cell-mediated metastatic renal cell carcinoma treatment called CreaVax-RCC by JW CreaGene and the liver cancer therapeutics called Immuncell-LC by Innocell (Innocell was acquired by Green Cross in 2012). Unlike somatic cells, stem cells are present in small amounts around specific body parts, and can differentiate into various types of cells such as muscles and nerves with infinite proliferation. Because of these characteristics, stem cells are emerging as important cell therapeutic ingredient in the regenerative medicine (Figure 9-8).

In a stem cell therapy, an adult stem cell therapeutics and an embryonic stem cell therapeutics exist. An adult stem cell can be classified into a bone marrow-derived stem cell, an umbilical cord blood-derived stem cell, and an adipose tissue-derived stem cell. An embryonic stem cell can be also called as a pluripotent stem cell, and it has the capability of differentiating into almost any types of cells in the body. Due to its essential nature, an embryonic stem cell has safety issues of bioethics and the rise of cancerous cells. Because of the immune rejection problem when a stem cell therapy can introduce to a patient, a customized genetically modified pluripotent stem cell was developed. These are called a cloned embryonic stem cell and an induced pluripotent stem cell (iPS). The cloned embryonic stem cell is obtained from an embryo by replacing the somatic cell and the oocyte nuclei, and the iPS cell can be obtained by inversely differentiating somatic cells by gene manipulation to reprogram back into an embryonic-like pluripotent state. The stem cells developed with these methods are actively under research because they do not have bioethical problems, but solution against lack of genetic stability is much needed. The use of adult stem cells is a lot safer than other stem cells, therefore, its related research and clinical trials are much more active.

South Korea also released some stem cell therapeutics in 2012 and got approved as the world's first stem cell therapeutics products. A myocardial infarction therapeutics called Hearticellgram-AMI by the FCB-Pharmicell Inc., a cartilage cell therapeutics called Cartistem by Medipost Co., Ltd., and the Crohn's disease fistula treatment called Cupistem by Antrogen are such products. As some Canadian products such as Prochymal as a treatment for Graft-vs-Host disease (GvHD) by the Osiris Therapeutics got approved as the first therapeutics for GvHD too, the use of stem cell therapeutics started to become more commercialized. GvHD is a serious complication that can occur after an allogenic stem cell transplantation as a donor's immune system starts to attack the recipient when the recipient's immune system becomes weaker.

The original technology being developed during the development of the new drug using stem cells will replace the preclinical tests such as animal experiments, and the clinical trials are expected to be conducted efficiently in a short period.

Overall, stability, efficacy and quality should be taken into consid-

Figure 9-9. In Vitro Diagnostics(IVD)

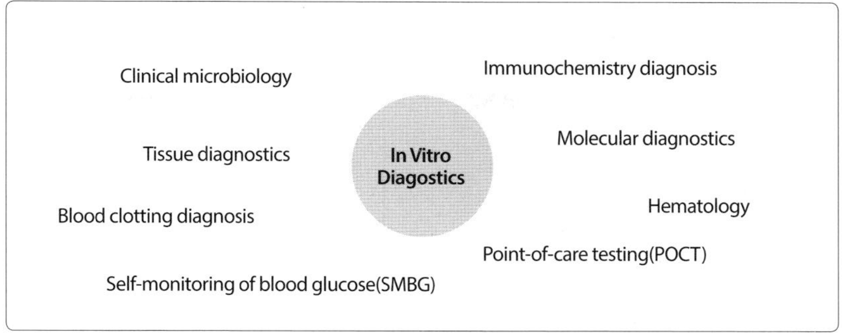

eration to develop highly valued cell treatments. However, the quality maintenance is challenging because cell treatment drugs are living biopharmaceuticals and have a high possibility of being contaminated by mircoorganisms.

Normally, blood or urine samples are collected during an individual's health check-ups, and are used to determine the health status, disease status or treatment effect, or as a prevention by using reagents or devices. These reagents and diagnostic devices used for diagnosis is mainly called an In Vitro Diagnostics (IVD) (Figure 9-9). The IVDs can be divided into multiple types on the basis of their detailed technology – the immunochemistry diagnosis using antigen-antibody reaction; the self-monitoring of blood glucose (SMBG) for personal diabetes management; the point-of-care testing (POCT) at the time and place of patient care; the molecular diagnostics for pathogen or cell DNA/RNA analysis; hematology to analyze white blood cells, red blood cells, platelets, and hemoglobin; clinical microbiology that determines the type and concentration of antibiotics to be injected by diagnosing the type of pathogenic infections; tissue diagnostics that diagnose based on normal cell tissue, cancer cell tissue, pathogen-infected cell tissue, etc.; hemostasis is a blood clotting diagnosis that diagnose the hemostatic disease, platelet disorders, and autoimmune conditions.

Based on the statistics released by the Frost & Sullivan in 2012, the followings are the list of orders of the global IVD sales. The immunochemical diagnostics shares about 35% of the market, SMBG market is about 20%, POCT is about 12%, the molecular diagnostics shares about 9%, hematology shares 7%, both clinical microbiology and tissue diagnostics market share were 6%, and hemostasis was about 3%.

According to the 2014 statistics, about 25% of the global IVD market is occupied by Roche Holding AG. Other than that, companies such as Siemens, Abbott Laboratories, Johnson & Johnson, and Danaher Corporation each hold about 10% of the shared market, and so many other companies are under competition. In Korea, most of the products are imported from other countries except some IVDs related to urinalysis, immunoassay, and molecular diagnostics. Out of all, mostly the molecular diagnostics-related products have been developed and produced by various domestic companies such as Osang Healthcare Co., Ltd., I Sens Inc., Macrogen, iNtRON Biotechnology, NanoEntek Inc., Seegene, Access Bio, and ViroMed Co., Ltd..

The in vitro diagnostic device market is gradually expanding as the older population has been growing all around the world and the frequency of the in vitro diagnostic device use has been increasing. Though the world market in 2014 is about 50 billion dollars and the Korean domestic market is about 600 million dollars, the revenue is continuously increasing. Looking at the revenue itself and the technological skills, the competition between the Korean in vitro diagnostic device businesses is very weak compared to other countries. However, a good result can be expected if Korea establishes a system for cooperative research and development in the related fields such as biotechnology, IT, or nanotecnnology and strives continuously since the development of in vitro diagnostic devices is basically based on fusion technology. As this chapter briefly reviewed on the biomedical field, we will look at the biochemical industry field using biomass in the next chapter.

Chapter 10

Future of Biomass and Industrial Biotechnology

Over the past millions of years, biomass such as cadavers of animals and plants were buried underground and chemically converted into fossil fuels such as coal, oil, and natural gas under high temperature and pressure. Since the Industrial Revolution, human beings started using the fossil fuels which caused a rapid growth and development of petroleum-based petrochemical industry, and it had a great impact on many other industries. The fossil fuels are irreproducible energy because they have limitations on the quantity and economically while the world energy consumption has been increasing. Therefore, the fossil fuels will be completely depleted someday in the future, and many nations are focusing on developing renewable energy or alternative energy. The types of renewable energy classified by the Energy Management Corporation of Korea are solar light, solar heat, wind

Figure 10-1. Different types of renewable energy

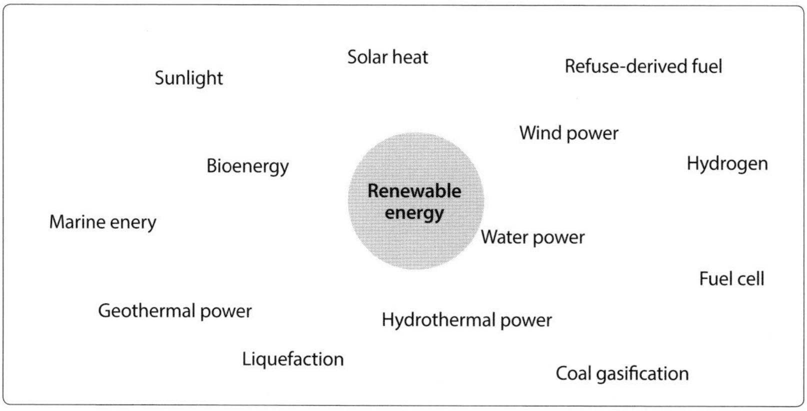

energy, bioenergy, waste energy, marine energy, hydropower, geothermal, hydrothermal, fuel cell, hydrogen, coal gasification and liquefaction. As many nations faced two times of the oil shock in the 1970s, they have been putting much effort to secure diverse energy sources rather than depending on the Middle East countries for oil supply (Figure 10-1).

Other than it being an irreproducible energy, using fossil fuels as a source of energy can also cause environmental problems such as global warming, acid rain, and respiratory diseases by producing carbon dioxide (CO_2), nitrogen oxides (NO_x), and sulfur oxides (SO_x). As environmental problems have arisen as important issues, use of bioenergy has attracted much interest as a way to reduce pollution from the exhaust gas produced from gasoline or diesel that is currently being used as the fuel for existing automobiles. In here, bioenergy refers to the energy produced from biomass. Biomass has the strength of being renewable, infinitely producible, and barely has any pollutant emission (Figure 10-2). Biomass generally refers to the total amount of organisms such as animals, plants, microorganisms, and other organic residues that can be recycled and these are the important living resources for the future. As mentioned briefly in Chapter 2, biomass can be classified into first, second, third, and fourth generations. First generation of biomass includes grains, second generation is wood biomass, third generation is algae, and fourth generation is organic residues. The primary reason why the scientists first started to show interest in biomass was because of bioenergy, but now they are mostly focused on the production of various products from renewable biomass. In particular, the white biotechnology field is developing for the process of producing various final products, intermediate products, and raw materials such as bioenergy, functional foods, medicines, fine chemical products, and biodegradable polymers using biomass through

Figure10-2. Cycle of renewable biomass

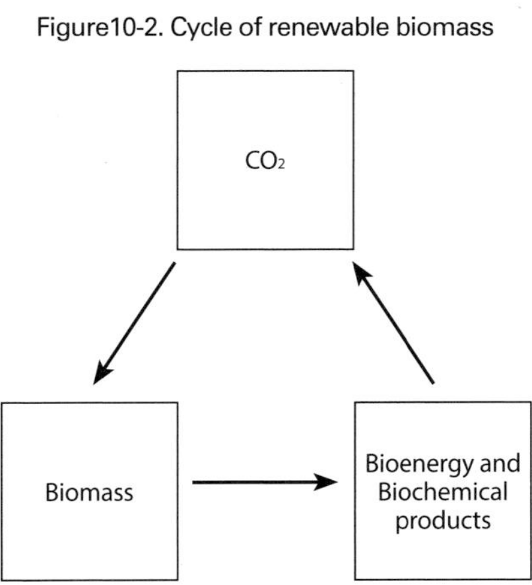

the chemical or biological conversions. This process is called biorefinery. The term biorefinery comes from a refinery meant to refine crude oil, and the chemical industries and chemical products actually started to bloom as the refinery was introduced (Reference Figure 2-9).

An industrial biotechnology refers to the production of biotechnological products by fermentation process using microorganisms or bioconversion reactions using enzymes. Currently, new attempts have been made to efficiently produce products that have been manufactured through chemical synthesis or products that could not be produced due to lack of technology, by introducing a technology for new strain development such as synthetic biology or metabolic engineering. The most important concept of biorefinery is that various biomass can be used continuously without intervening the environment. For example, biomass grows by synthesizing carbohydrates and using them as nutrients as they go through photosynthesis using carbon dioxide produced from factories and the transportation fuels. The National Renewable Energy Laboratory (NREL, USA), one of the most promising laboratories related to biomass, divided the concept of a biorefinery into two different platforms – a sugar platform and a syngas platform. In here, the sugar platform produces products by breaking down the biomass into different types of component sugars for fermentation based on the bioconversion reaction. The syngas platform is based on the thermochemical conversion reaction. As biomass is converted into syngas, it is also converted into the by-product. In other words, the core technologies are developed and the final products are produced based on these platforms. Though the NREL classified a biorefinery into two different platforms, various types of platforms can be introduced as the technology develops and depending on the types of raw materials derived from biomass. As an example, a core technology using lipids that contain many carbons and hydrogens or biogas as the raw material can arise in the field.

Now, let's look into the characteristics of different types of biomass (Reference Figure 2-7).

The first generation biomass includes grains such as corn, wheat, barley, cassava, and potatoes. Starch can be obtained from these food crops, be converted into saccharides, and variety of products can be produced.

However, since these food crops are used as the main food resources, the cost is expensive. Therefore, products can be obtained from the sugar crops such as sugarcane, sugarbeet, or sweet sorghum. The situation is quite complicated due to the entangled political relationship between the Organization of Petroleum Exporting Countries (OPEC) and shale gas and biofuel supply countries. The energy war between the politically empowered countries such as United States, Russia, and China is especially competitive. Currently, the United States is producing bioethanol from corn starch, Brazil is producing bioethanol from sugarcane, and other countries with abundant available resources are producing bioethanol from various raw materials. On the other hand, using these types of biomass to produce biofuels is being internationally criticized due to the global food shortage.

Therefore, the research related to the conversion of the wood-type, marine, and organic biomass residues into the biofuel is much popular compared to the biofuel conversion of foods such as grains.

The second generation biomass is mainly wood biomass which includes energy crops such as poplar, willow and reed, and agricultural byproducts such as rice straw, barley straw, wheat straw, rice bran, corn stover, sugarcane bagasse, oil palm empty fruit bunch (OPEFB), and municipal or forest residues such as newspaper, pulp and paper residues, waste wood. Wood biomass is made of three ingredients – cellulose, hemicellulose, and lignin. The composition ratio of these three components that makes up the fibrous biomass varies depending on the type of biomass, but is generally about 4:3:3. Cellulose and hemicellulose are usually broken down to 6 carbon glucose and 5 carbon xylose by chemical pretreatment using acid and alkaline, and by saccharification using enzymes. Currently, new pretreatment

Figure 10-3. Composition of lignocellulosic biomass

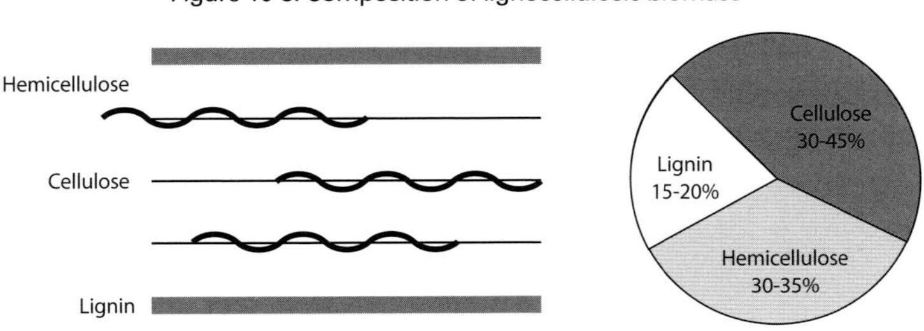

Figure 10-4. Schematics for the production of biomaterials, bioenergy, and biochemicals from lignocellulosic biomass

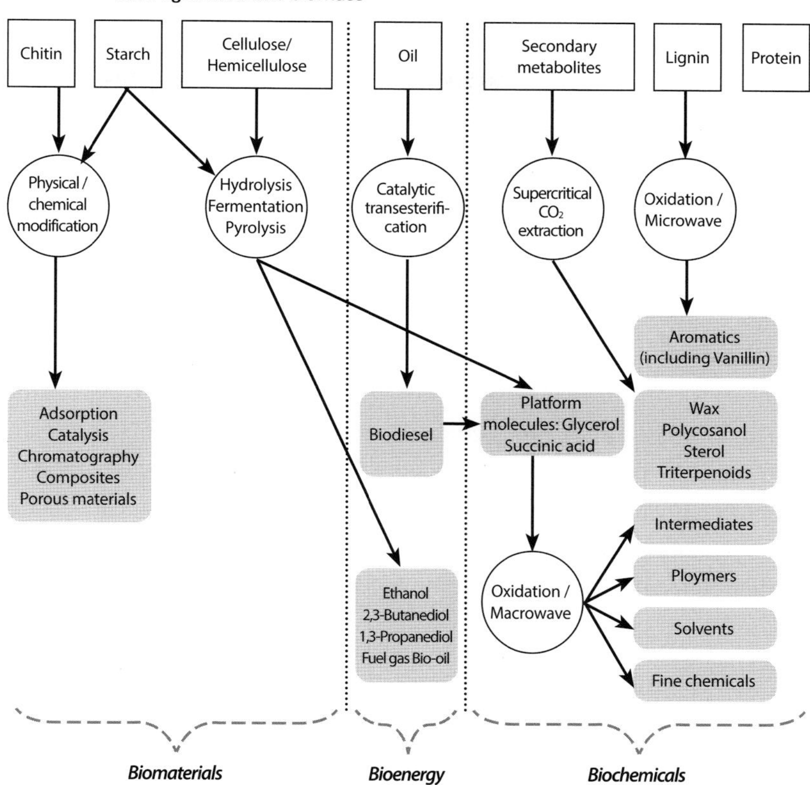

methods are also under research. Lignin is generally physiochemically bonded with cellulose and hemicellulose and is composed of fat-soluble phenolic polymers which are complex aromatic polymers. Therefore, the saccharification is easier as the content of lignin decreases. In the recent years, studies have been made to establish a sugar platform for efficient degradation of lignocellulosic biomass into glucose and xylose. Also as a byproduct, lignin is no longer considered as a pollutant, but can be used as the base material for epoxy resins and high calorific fuels. When cellulose is broken down, glucose is mainly produced but hemicellulose can produce more saccharides other than xylose such as galactose, mannose, arabinose, and rhamnose (Figure 10-3).

A variety of studies are being conducted to produce bioethanol for fuel and various high value-added products by developing the economical process that can decompose these types of materials more efficiently. For these lignocellulosic biomasses to be efficiently degraded and for useful products to be produced, research and development on pretreatment, saccharification, fermentation, and new strains related to fermentation and hydrolytic enzyme are critical (Figure 10-4).

A general pretreatment method includes physical, chemical, and biological methods (Figure 10-5). First of all, physical methods such as grinding or milling, steam explosion, or irradiation must be performed to break down the lignocellulosic biomass into small fragments. Once the biomass is broken down into small fragments, the surface area increases and the enzymes can be easily contacted with the biomass, which increases the efficiency of pretreatment. Using steam explosion can also increase the pretreatment efficiency by increasing the surface area and decreasing the crystalline structure of the lignocellulosic substance. The irradiation method can reduce the crystalline structure of the lignocellulosic substance

Figure 10-5. Pretreatment methods for lignocellulosic biomass

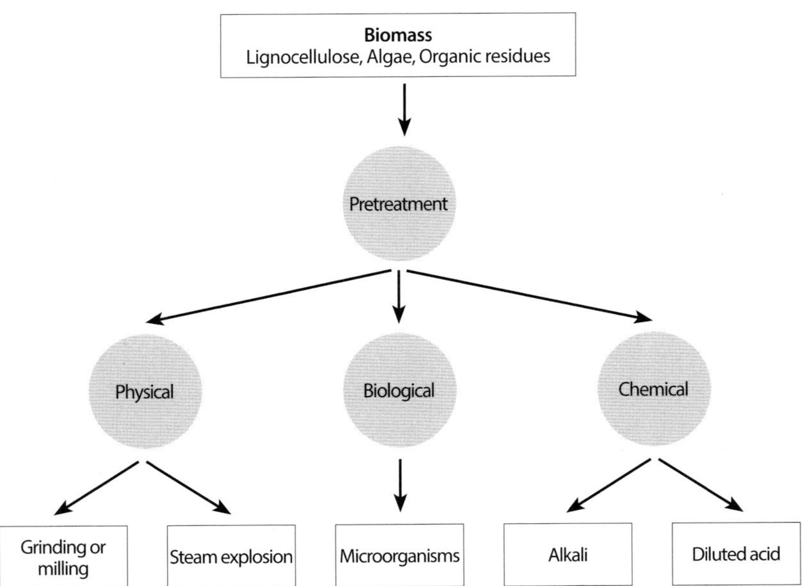

by beaming the proton beam, an ion beam, a γ-ray, or electrons. Currently, the steam explosion with addition of acid or alkaline solvent is being widely used, but has the weakness of producing high quantity of inhibitors after the pretreatment.

The chemical method normally uses weak acids or alkalines, supercritical fluids, and ionic liquids. In here, the superciritical fluid is a type of fluid that has traits of both gas and liquid and the distinct liquid and gas phases do not exist once the temperature and pressure are above its critical point. Therefore, supercritical fluid has the advantage of having a good diffusion property as gas and has good solubility as liquid. The only disadvantage is that the initial equipment set-up cost is quite expensive. Ionic liquid is a salt which the positive and negative ions are poorly coordinated due to the asymmetry of the sizes, and the ions cannot form crystals and stays in a liquid state under 100℃. Although it is known as an environmentally friendly solvent, the cost is expensive to use. If the pretreatment is performed under an appropriate concentration of acids and optimum temperature, xylose can be collected and some of the lignin is also separated. In other words, the cellulosic tissues become tender as the lignocellulosic structure is destructed by removing lignin. The method of alkaline use is to remove lignin by immersing biomass into ammonia. Ammonia can be recollected and recycled. Other than that, other methods use a two-step combined process of pretreatment method where the diluted acid is used in the first step, then the water-soluble ammonia is used as the second step.

Biological methods mainly use fungi that secrete enzymes that can break down lignin and this method is only used in special occasions because the time takes longer. Therefore, many scientists have been studying to find stronger lignin-degrading enzymes to shorten the break-down time.

As xylose and lignin are collected, they are used in appropriate consequences, and rest of them are saccharified by the cellulose degrading enzymes to collect glucose. As mentioned briefly before, this process of converting cellulose into glucose after pretreatment is called saccharification. In the past, efforts have been made to obtain glucose through acid treatment, but this process needed much optimization such as high acid concentration or low acid concentration with high temperature. During this process, the equipment that does not easily get rusty by acid needed to be

Figure 10-6. Various strains producing the saccharifying enzymes which degrade the lignocellulosic biomass

Fungi	*Bacteria*
Acremonium celluloyticus	*Clostridium thermocellum*
Aspergillus acculeatus	*Ruminococcus albus*
Aspergillus fumigatus	*Streptomyces sp.*
Aspergillus niger	*Thermoactinonyxes sp.*
Fusarium silani	*Thermomonospora curvata*
Irpex lasteus	*Bacillus substillis*
Penicillium funmiculosum	*Bacillus pumilis*
Phanerochaete	*Bacillus mycoides*
Schizophyllum commune	*Pseudomonas fluorescens*
Sporotrichum cellulophilum	*Serratia marscens*
Talaromyces emersinii	*Cellulomonas xylanilytica*
Thielavia terrestris	
Trichoderma koningii	

used, and also had a disadvantage of produced glucose being transformed into various byproducts. As the cost of enzymes have decreased, and as more active and stable enzymes have been introduced now, obtaining glucose with the use of enzymatic method is much preferred. The study of finding better strains with higher enzymatic activity and stability is still ongoing and based on this, the fixed production cost can be lowered by developing the cocktail enzymes (Figure 10-6).

Currently, the strains that produce one of the strongest cellulose-degrading enzymes are the mutant strains from the strain called *Trichoderma reesei* developed by the U.S. Army Natick Laboratory. These mutant strains are effective in breaking down cellulose, but cannot break down hemicellulose nor lignin. In particular, these strains have a major disadvantage of having a very low β-glucosidase activity which is an enzyme that breaks down two-bonded glucoses called cellobiose into single glucose. The strong enzymes produced by these strains mainly exist as the complex of endoglucanase and exoglucanase. While endoglucanase randomly cleaves internal glycosidic bonds in a glucose polymer, exoglucanase cleaves the outside bonds of the cellulose chain. The endoglucanase-exoglucanase enzyme complex generally breaks down cellulose into cellobiose and glucose. However, these products can act as inhibitors of

Figure 10-7. Cellulose degradation by cellulose-degrading enzyme systems

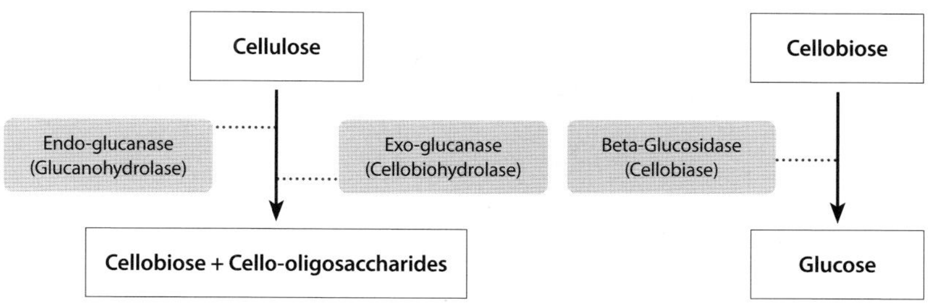

this enzymatic reaction. Cellobiose must be immediately converted to glucose by β-glucosidase. β-Glucosidase can be used after being obtained from different types of strains. The enzyme reaction can also be inhibited when the concentration of glucose is too high. Therefore, many scientists are still trying to develop an effective process by producing the cocktail enzymes which many different cellulose-degrading enzymes produced from different strains are all mixed together for the highest efficiency (Figure 10-7).

The most important factors in the enzymatic hydrolysis of cellulose are temperature, pH, amount of enzymes being added, concentration of the substrate or product, and concentration of the inhibitor. The productivity increases only when the conversion rate and yield are maximized by optimizing all these factors.

The South Korea's land very much lacks with natural resource availability including the wood biomass. Therefore, much effort has been made to plant more rape plant or reed and to use the residues as biomass. With local characteristics, specific biomass such as barley straw and rice straw are being well-harvested which helps with studies of bringing higher production of bioethanol or chemical products. Rape plant is mainly used as canola oil in food or can produce biodiesel. The rest ingredients of the rape plant are lignocellulose, therefore, can be made as various products after the pretreatment and saccharification. The residues are handled under anaerobic environment which can produce methane gas. The reed grows near the waterfront and has the advantage of absorbing heavy metals to help the environment. It can also be used as the lignocellulosic biomass when harvested. With less possession of biomass, much effort is needed to

Figure 10-8. Various kinds of algal biomass

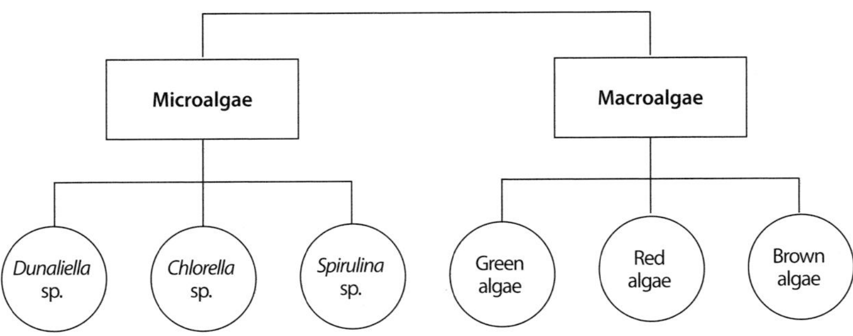

obtain and retain the biomass. It is also important to develop and export potential core technologies using the biomass acquired from other countries.

The third generation biomass, algae, is widely distributed in seawater and freshwater and can be classified as microalgae and macroalgae. Much like plants, algae go through photosynthesis by using the sunlight, water, and carbon dioxide. Microalgae are normally seen under the microscope, and are also called as the plant plankton. Some popular species are Chlorella and Spirulina. Macroalgae are generally multicellular, easily visible with the naked eye, and are also called as the seaweeds. Green laver, seaweed, agar are such types. Macroalgae can be divided into different types depending on different colors – green algae, red algae, and brown algae (Figure 10-8).

Algae is normally considered as an important organism in food, medicine, bioenergy or environmental industries because it contains various substances such as carbohydrates, polysaccharides, lipids, and proteins, and has the ability to produce functional materials. Macroalgae vary depending on the type, but most of them contain carbohydrates such as cellulose, pectin, xylan, or mannan, and polysaccharides such as alginic acid, carrageenan, and agar. Microalgae contain various ratio of lipids, carbohydrates, polysaccharides, and proteins.

To use macroalgae and microalgae efficiently, acids or alkaline or appropriate enzymes can be used like pretreatment, when using wood biomass. Since microalgae contain lipids, biodiesel can be produced by first extracting lipids while cellulose is converted to glucose and galactan from agar is converted to galactose. Various chemical products and bio-

Figure 10-9. Application of algal biomass

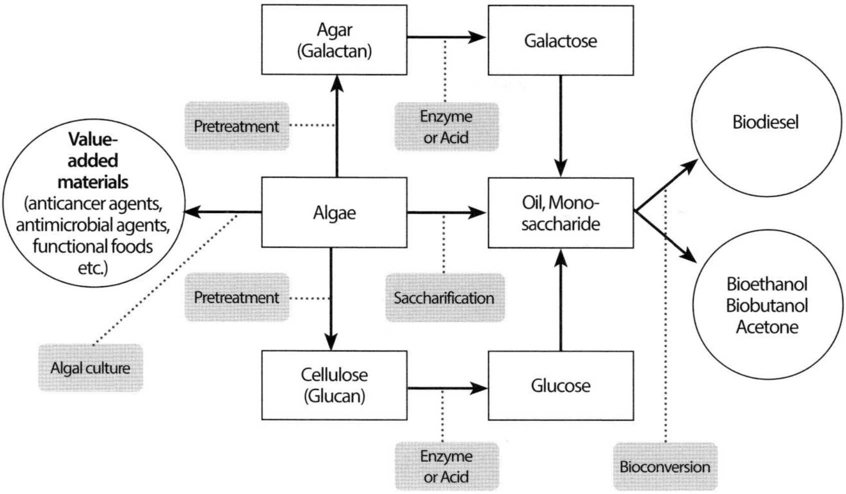

ethanol can be produced through the conversion of glucose and galactose.

The recent studies are continuing to develop a strain capable of producing maximum amount of lipids and developing the mass production of biodiesel to ultimately produce the jet fuel from microalgae. Not only that, studies on microalgae culture using CO_2 that is contained within the flue gas from factories and electric power supply facilities are also actively under research. Plus, the effort to produce high value-added products during the microalgal culture has also been made in various ways (Figure 10-9).

The fourth generation biomass is organic residues, which technically contain the second generation biomass such as agricultural byproducts, waste woods, newspapers, and waste papers but these are mainly considered as wood biomass. High quantity of organic residues is made from various industries, especially from the food industry (Figure 10-10).

Various residues from food industry includes molasses that can be used after sugar is collected from the sugarcane, corn steep liquor that can be used after starch is collected from corn, whey that is a byproduct of cheese, residues of ramen noodles, coffee, and herbal medicines. For example, molasses contains over 50% of sugar which enables it to be used as the substrate of fermentation. Corn steep liquor has high content

Figure 10-10. Various kinds of organic residues and their application

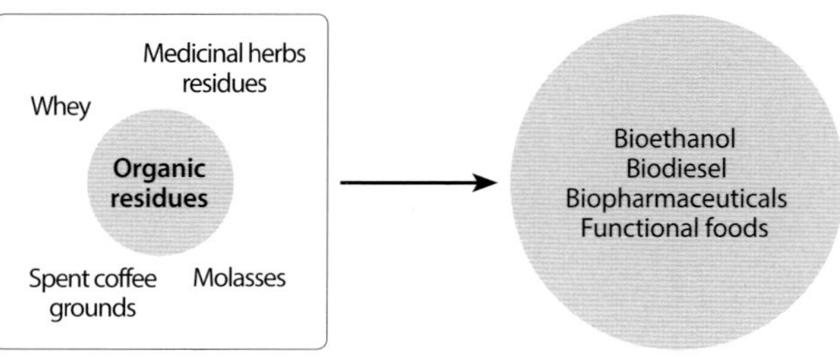

of amino acids and nitrogen compounds which enables it to be used as a nitrogen source for the fermentation medium. Whey contains high percentage of lactose which can be used as a bioconversion substrate to produce lactulose, which is used as a functional food or pharmaceutical. The ramen noodle remnant contains starch and lipid that can produce bioethanol and biodiesel. Although most of the residues from the food industry are being used in the cheap animal feed currently, they can be used as good energy resources if more studies are done on the details of the ingredients and the reassurance on possibilities of producing high value-added products.

So far, we looked into the types and characteristics of biomass. Biomass is made of substances such as functional materials, starch, cellulose, hemicellulose, lignin, lipid, and protein. Therefore, various products such as bioenergy, biopolymers, and biochemical products can be produced efficiently by using the substances from biomass.

Now, let's talk about the process of producing some final products from biomass.

We will first talk about bioenergy such as bioethanol and biodiesel.

Biodiesel and bioethanol are two important commercialized transportation bioenergy. Biodiesel is mainly made from the plant oils or lipids, and bioethanol is made from grains and wood biomass. Depending on the country, some nations use each 100% of biodiesel (BD100) and bioethanol (BE100) and, in some other nations, a certain ratio of biodiesel or bioethanol is mixed with the diesel or gasoline respectively for the use. For example, BD10 means 90% diesel and 10% biodiesel are mixed together, and BE10 means 90% of gasoline and 10% of bioethanol are mixed together. The fuel that gasoline and bioethanol are mixed together is called gasohol. The followings are the advantages of mixing these fuels into the automobile fuel. In the case of biodiesel, nitrogen oxides may increase slightly but the level of carbon monoxide, hydrocarbon, and dust are significantly reduced. This fuel has high cetane number and high oxygen content, therefore, can be mixed and used with diesel without the transformation of the car engine. Bioethanol can reduce emission of carbon monoxide and hydrocarbons and increases the burn rate. Also, increased octane number can contribute to prevention of global warming.

When analyzing the carbon and energy flow during the production and use of bioethanol fuel from biomass, a large amount of biomass can be synthesized and regenerated every year by photosynthesis using atmospheric carbon dioxide. Therefore, if this is converted to bioethanol, the carbon cycle becomes balanced.

Since long time ago, Brazil has been producing bioethanol from sugarcane as a use of automobile fuel. This shows ethanol is a great alternative energy for gasoline. A fuel bioethanol as an alternative energy can be converted from the wood biomass and the agricultural wastes. In the recent years, many countries have been developing fast-growing hybrid plants such as hybrids of poplar, willow, reed, and grass to use them in the biorefinery processes for the biochemical industries.

As mentioned earlier, xylose can be converted into bioethanol by using yeast that can ferment xylose itself or using yeast that can ferment both xylose and glucose. This can be performed after xylose or glucose is produced through biomass pretreatment and saccharification. The yeast that can ferment both glucose and xylose at the same time are still under development and have not been industrialized yet. The most traditional way to ferment glucose into bioethanol is to use the separate hydrolysis and fermentation (SHF). SHF uses a method of breaking down cellulose into glucose by using the cellulose-degrading enzymes and then let glucose go under fermentation with a certain yeast strain. A drawback to this method is that high concentration of yeast is needed due to the strong inhibition effect of cellobiose and glucose, and two bioreactors are needed during the process which makes the overall production cost quite expensive. To overcome this problem, another method called the simultaneous saccharification and fermentation (SSF) is also mainly used which makes the

Figure 10-11. Bioprocess for bioethanol production

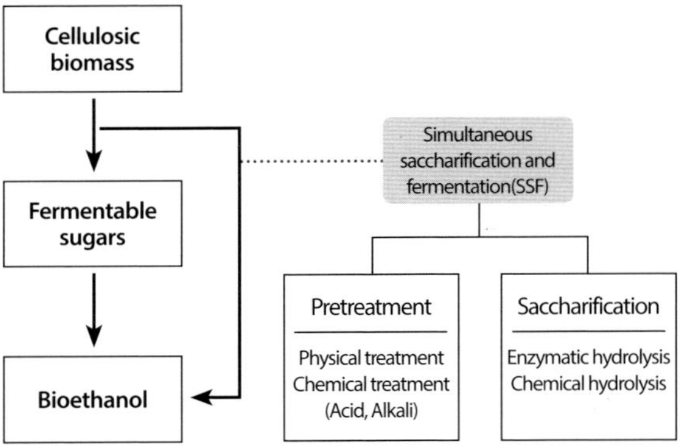

saccharification by enzymes and fermentation by yeast occur within one bioreactor at the same time. Unlike SHF, SSF enables the produced glucose to be directly converted into bioethanol which reduces the accumulation of glucose and its inhibition effect towards the product. This increases the overall productivity because the input of the cellulose-degrading enzymes is significantly lowered. In general, the optimal reaction temperature for the cellulolytic enzymes is about 50 °C, and the optimum temperature for bioethanol fermentation by yeast is 30-35 °C. Therefore, the optimum condition between these two temperatures should be established and is more productive to use the thermotolerant yeast (Figure 10-11).

Studies on using strains that directly converts biomass to bioethanol after the wood biomass pretreatment under the anaerobic condition also exist, but this case is not available yet due to the low productivity.

The recent studies have been focusing on developing yeast that can ferment glucose and xylose at the same time. The most well-known yeast is called *Saccharomyces cerevisiae* that can only ferment a 6-carbon glucose. Though the scientists have developed a yeast that can also ferment a 5-carbon xylose through gene modification, it is not being utilized due to the long fermentation period and low bioethanol productivity. Therefore, much research is being conducted with *Pichia stipitis* also, because of its lower resistance towards bioethanol than *S. cerevisiae* and its diauxic aspect of using glucose first and then xylose as substrates. With the characteristics of *P. stipitis* using glucose first, the scientists should develop and

Figure 10-12. Biomass for biodiesel production

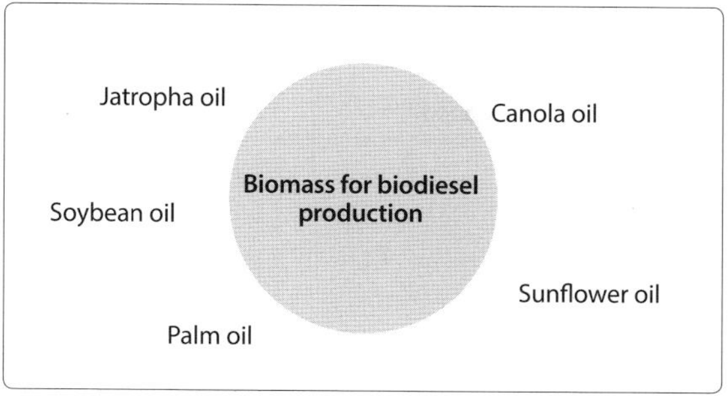

find a way to manipulate the process by inhibiting the use of glucose to a certain degree so that the yeast uses glucose and xylose at the same time and for it to have less resistance towards bioethanol. Another study shows an ability of genetically modified *S. cerevisiae* to play SSF by expressing the cellulose-degrading enzymes onto the yeast cell surface.

Currently, the process for bioethanol production from biomass is focusing on the production of enzymes, development of strains that produce bioethanol, and development of pretreatment technologies.

Biodiesel is normally produced by transesterification of lipids that are obtained from vegetable oils such as palm oil, soybean oil, canola oil and jatropha oil or from algae, and is composed of a mixture of fatty acid esters of 16-20 carbon atoms. Among vegetable oils, the highest productivity per unit area is obtained from palm oil. Comparing with the diesel, production cost is cheaper when using waste oil and is slightly higher when using the canola oil or soybean oil. The important factors to consider when producing biodiesel are the concentration of free fatty acids, water content, oxidation degree, and viscosity (Figure 10-12).

The method of producing biodiesel can be largely classified into the chemical method and enzymatic method, and the supercritical fluids can be used in each method (Figure 10-13). Biodiesel is produced in the presence of catalyst. As triglyceride and alcohol (methanol or ethanol) react together, the biodiesel such as fatty acid methylester (FAME) or fatty acid ethylester (FAEE) is produced, and glycerol is produced as a byproduct.

When using the chemical method, an acid or a base is mainly used as

Figure 10-13. Mechanism for biodiesel production

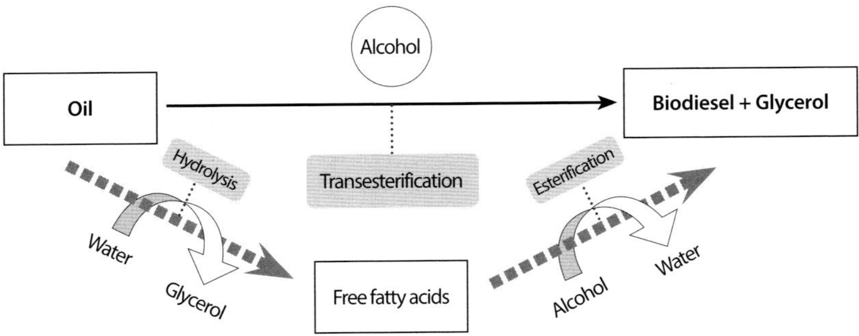

a catalyst, and much research on developing a solid catalyst for efficient and economical production of biodiesel is underway. Using a solid catalyst has an advantage of producing high purity glycerol. Normally, an acid catalyst is used when the concentration of free fatty acid contained in the raw material exceeds 2%, and the reaction is carried out at a pressure of 80 bar and a temperature of 250 ℃ for 2-4 hours. When the free fatty acid contained in the raw material is less than 2%, a base catalyst is used at a pressure of 9 bar and a temperature of 100 ℃ for 10-30 minutes. Therefore, the two-step process of converting the free fatty acid into biodiesel by using an acid catalyst first and then using a base catalyst is used when the free fatty acid concentration is high like the cooking oil waste.

An enzymatic method generally uses lipase as an enzyme catalyst, and the reaction does not depend on the free fatty acid concentration contained in the raw material. Since the reaction condition does not require high temperature and pressure and any solvents, this method is considered as environmentally friendly. The process is also carried out very simply because it does not require a complicated separation process. The only disadvantage is that the reaction rate is slow which takes about 24-48 hours to complete, and the cost of enzymes is expensive. Therefore, the enzyme should be immobilized to improve stability and to enable recycling for a long period of time. During the reaction, the enzyme can be destroyed and lose its activity when it contacts with alcohol. Therefore, it is important to add small amounts of alcohol for multiple times so that alcohol does not affect the enzyme. Also, development of alcohol-resistant lipase is another challenge. In the enzyme reaction, triglycerides are first partially broken down through hydrolysis reaction, and then the esterification reaction with alcohol occurs. During this reaction, the reaction rate becomes slower due to the migration time of the acyl group. The reaction rate must become faster in order for the enzymatic method to compete against the chemical method, and this can be resolved by using a different type of lipase. Many different types of lipases are produced by different microorganisms such as 1,3-specific lipase, 1,2-specific lipase, and nonspecific lipase. The reaction rate stays slow when only one kind of lipase is used during the process. However, the reaction time can be shortened when two different types of lipases are mixed and used together. They can be used together by being separately immobilized in two different carriers or can be immobilized together in one carrier. When the immobilized lipase is filled into the packed-bed bioreactor and introduced with substrates, biodiesel can be produced continuously. The current studies are developing a way to produce biodiesel by genetically expressing lipase onto the surface of yeast or *E. coli* (Figure 10-14).

As explained above, glycerol is produced as a byproduct when producing biodiesel through the chemical or enzymatic reactions. This is

Figure 10-14. Lipase-producing strains

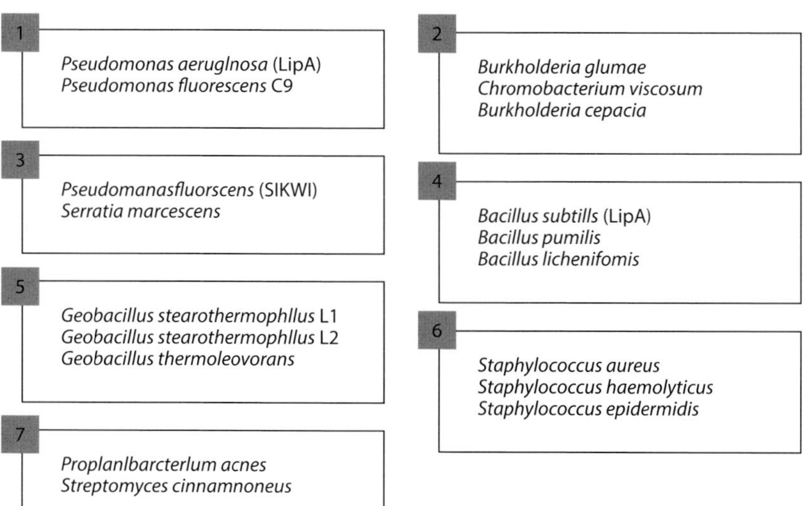

called crude glycerol. Many studies are being conducted to convert this crude glycerol into the high value-added products. Though it is a popular ingredient for the cosmetics industry, the cost has gotten cheaper due to its increased quantity with the increase of biodiesel production. Therefore, much effort is being made to convert glycerol into various materials by using chemical, biological, and enzymatic methods. For example, conversion of glycerol can produce various bioenergy such as bioethanol, biobutanol, and biohydrogen, and can also produce polyester resin, anti-freeze material, surfactant, moisturizers, and propylene glycol as a basic raw material for other cosmetics. Materials such as 1,3-propanediol and 1,2-propanediol can also be produced that can be developed into anti-freeze, cosmetics, medicine, and polyester, and glycerol carbonate can also be produced to be used in surfactants, electrolytes, and cosmetics (Figure 10-15).

Many businesses have been trying to produce an intermediate product called 2,3-butanediol from the saccharide derived from a biomass to develop various plastics and pesticides. Since 2012, companies such as Coca-Cola, Ford, Heinz, and some other companies have joined the consortium to develop a process to produce terephthalic acid (TPA), a raw material for various polyesters, from biomass-derived saccharides and lignin for commercialization. Currently, they are producing TPA from the fossil fuels, synthesizing polyester, and producing beverages, films, and fibers.

Figure 10-15. Biochemical products produced from glycerol

Glycerol carbonate
Surfactants, Electolytes, Cosmetics

Propylene glycol
Polyester resin, Antifreeze, Surfactants, Moisturizer, Cosmetics

Biochemical products produced from glycerol

Bioenergy
Bioethanol
Biobutanol
Biohydrogen

1,3-Propanediol
1,2-Propanediol
Antifreeze, Cosmetics, Biopharmaceuticals, Polyester

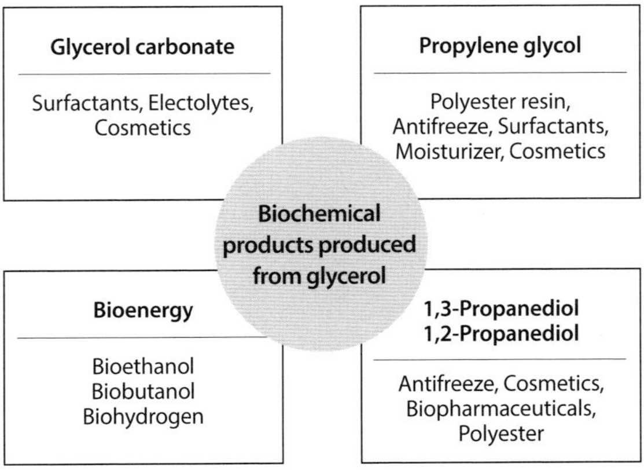

A good example of food developed from xylose is the xylitol. Xylitol is mainly used in food preservatives, beverages, chewing gums, and toothpastes. When yeast uses xylose, xylose is converted to xylitol, then to xylulose, and xylulose goes under the glycolysis pathway to produce energy. When the metabolism step from xylitol to xylulose is genetically inhibited, the related enzymes cannot work in the system which makes xylitol as the final product. The glycemic index of xylitol is only 8. In here, the glycemic index is a relative ranking of carbohydrates in foods from 0-100 according to how they affect blood glucose levels, and has glucose as the standard of 100.

The example of the pharmaceutical field would be the production of antibiotics called Cephalosporin C (CPC) through fermentation process. CPC is then chemically or enzymatically converted into the intermediate product called 7-aminocephalosporanic acid (7-ACA) and is being introduced to the antibiotic markets. From 7-ACA, various products of Cephalosporins are being produced through chemical synthesis. To produce CPC, glucose and methionine are required as substrates. However, CPC can also be produced from the biomass-derived materials such as xylose and glycerol.

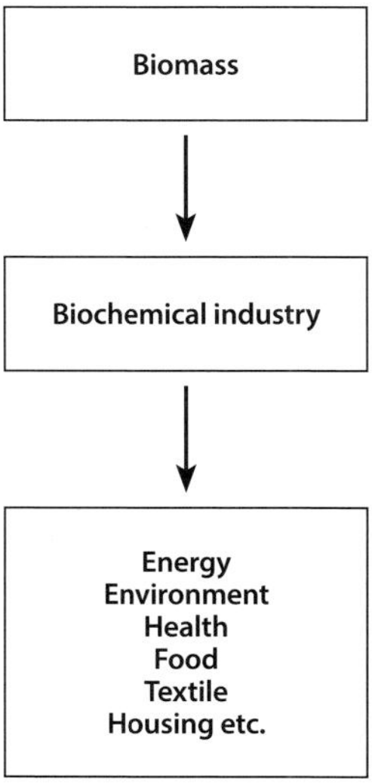

To summarize the whole chapter 10, the building blocks, intermediate products, and final products can be produced through the cell metabolism, biotransformation by cells or enzymes, and chemical synthesis by using starch, cellulose, hemicellulose, lignin, lipid, protein, and saccharide that can be obtained by various biomass. By going through these processes, various products related to energy, environment, health care, food, textile, and housing can be produced as a part of the biotechnology field (Figure 10-16). Nowadays, petroleum-based chemical industry is gradually evolving into biomass-based biotechnology industry.

Figure 10-16. Biomass and biochemical industry

Biomass
↓
Biochemical industry
↓
Energy
Environment
Health
Food
Textile
Housing etc.

References

Chapter 1

Barnum, S.R., Biotechnology – An Introduction, 2nd Ed., Thomson, 2005.

Biotech. Policy Research Center(BPRC), BioINdustry, 99, Global biotechnology, 2015.

Duderstadt, J. J., A university for the 21st century, The university of Michigan, 2000.

Kang, S. W. and Kim S. W.(2001), Current status and prospect on biotechnology, Korean Industrial Chemistry News, 4(1), 34-44.

Korea Institute for Industrial Economics & Trade(KIET), Korea Industry Vision 2020, 2005.

MEST & BPRC, Biotechnology in Korea, 2011.

Ministry of Science, ICT and Future Planning(MSIP), Biotechnology 2013 & 2015.

MSIP, MOHW, MOE, ME, MAFRA, MOF, MOTIE, and MFDS, Bio-Vision 2016, 2014.

Oh T. K., Strategy and agenda for an innovation of national BT, STEPI Science and Technology Policy Forum., 2013.

Samsung Economic Research Institute(SERI), CEO Information, 667, Korea's 60-year plan, 2008.

Thieman, W. J. and Palladino, M.A., Introduction to Biotechnology, Pearson, 2004.

UNESCO Report, Engineering: issues, challenges and opportunities for development, 2010.

Chapter 2

Biotechnology Laboratory at Korea University, An unfamiliar story in laboratory, Commemorative retirement for Prof. Bang Won Ki, 2010.

Fleming, A.(1929), On the antibacterial action of cultures of a Penicillin, with special reference to their use in the isolation of B. influenza, The British Journal of Experimental Pathology, 10(3): 226-236.

Ha T., Trends in health functional foods, BioINpro, 15, 1-13, Biotech. Policy

Research Center (BPRC), 2015.
Institute for International Trade(IIT), Activation of Nagoya Protocol and its effect to industry, Trade Brief, 59, 1-8, 2014, Korea International Trade Association(KITA).
Jin T. E.(2015), Bioresource, Core material of bio-economy era, Science & Technology Policy, 20-25.
Kim, S. W.(2006), Organisms feed the future, Sisajournal, 80-81, 2007-06-05.
Kim, T. E. and Kim, S. W., Market for functional food industry and R & D for natural materials, BioINpro, 8, Biotech. Policy Research Center(BPRC), 2009.
Ministry of Science, ICT and Future Planning(MSIP), Biotechnology, 327-334, 2013.
MSIP, MOHW, MFDS, MAFRA, MOE, RDA, MOTIE, MOF, KFS, Enforcement plan for management of bioresources, 2015.
Perlman, D.(1999), Some problems on the new horizons of applied microbiology, Journal of Industrial Microbiology & Biotechnology, 22, 430-438.
The Korea Chosun Daily, "Ancient Future" Nature and Culture, Think of life and death, 2010-12-16.
Time, Bacteriologist ALEXANDER FLEMING, Mar. 29, 1999.
Yoon Y. M.(2014), Biomass utilization in Korea and its activation plan, World Agriculture, 162, 1-25, Korean Studies Information Service System(KISS).

Chapter 3

Brock, T.D. and Madigan, M.T., Biology of Microorganisms, 6th ed., Prentice-Hall, 1991.
Brown, T.A., Gene Cloning and DNA Analysis – An Introduction, 6th ed., Wiley-Blackwell, 2010.
Campbell, N.A., Reece, J.B., Urry, L.A., Cain, M.L., Wasserman, S.A., Minorsky, P.V., Jackson, R.B., Biology – A Global Approach, 10th ed., Pearson, 2015.
Enger, E.D., Kormelink, J.R., Ross, F.C. Smith, R.J., Concepts in Biology, 7th ed., Wm. C. Brown Communications, Inc., 1994.
Gibson, D.G., Glass, J.I., Lartigue, C., Noskov, V.N., Chuang, R-Y., Algire, M.A., Benders, G.A., Montague, M.G., Ma, L., Moodie, M.M., Merryman, C., Vashee, S., Krishnakumar, R., Assad-Garcia, N., Andrews-Pfannkoch, C., Denisova, E.A., Young, L., Qi, Z-Q., Segall-Shapiro, T., Calvey, C.H., Parmar, P.P., Huchison III, C.A., Smith, H.O., Venter, J.C.,(2010), Creation of a bacterial cell controlled by a chemically synthesized genome, Science,

329, 5987, 52-56.

Hutchison III, C.A., Chuang, R-Y, Noskov, V.N., Assad-Garcia, N., Deerinck, T.J., Ellisman, M.H., Gill, J., Kannan, K., Karas, B.J., Ma, L., Pelletier, J.F., Qi, Z-Q., Richter, A., Strychalski, E.A., Sun, L., Suzuki, Y., Tsvetanova, B., Wise, K.S., Smith, H.O., Galss, J.I., Merryman, C., Gibson, D.G., Venter, J.C.,(2016), Design and synthesis of a minimal bacterial genome, Science, 351, 6280, 1414-& aad6253-1 – 11.

Lane, N., Power, Sex, Suicide – Mitochondria and the Meaning of Life, Oxford University Press, 2005.

Lee, J.M., Biochemical Engineering, Prentice-Hall Inc., 1992.

Renneberg, R. and Berkling, V., Biotechnologie fur Einsteiger, 4. Auflage, Spektrum Akademischer Verlag, 2012.

Shuler M.L. and Kargi, F., Bioprocess Engineering – Basic Concepts, 2nd ed., Prentice-Hall Inc., 2002.

Stanbury, P.F. and Whitaker A., Hall, S.J., Principles of Fermentation, 2nd ed., Butterworth Heinemann, 1995.

Starr C. and Taggart H. Cell Biology and Genetics in Biology: The Unity and Diversity of Life, 8th ed., Wadsworth Publishing Company, 1998.

Chapter 4

Bloomfield, M.M. and Stephens, L.J., Chemistry and the Living Organism, 6th ed., John Wiley & Sons, Inc., 1996.

Buchholz, K., Biocatalysts and Enzyme Technology, Wiley-VCH, 2005.

Fessner, WD. Biocatalysis from Discovery to Application, Springer-Verlag, 2000.

Guisan, J.M., Immobilization of Enzymes and Cells, 3rd ed., Humana Press, 2013.

Kohler, V., Protein Design: methods and applications, 2nd ed., Humana Press, 2014.

Lee, J.M., Biochemical Engineering, Prentice-Hall Inc., 1992.

Minteer, S.D., Enzyme Stabilization and Immobilization: methods and protocols, Humana Press, 2011.

Palmer, T. Understanding Enzymes, 4th ed., Prentice Hall/Ellis Horwood Ltd., 1995.

Price, N.C. and Stevens, L., Fundamentals of Enzymology – The cell and molecular biology of catalytic proteins, 3rd ed., Oxford University Press, 2000.

Pundir, C.S., Enzyme Nanoparticles, William Andrew Publishing, 2015.

Purich, D., Enzyme Kinetics: Catalysis & Control, Elsevier, 2010.

Shuler M.L. and Kargi, F., Bioprocess Engineering – Basic Concepts, 2nd ed.,

Prentice-Hall Inc., 2002.
Smith, J.E., Biotechnology, 5th ed., Cambridge, 2009.
Walsh, G and Headon, D. Protein Biotechnology, 1st ed., John Wiley & Sons, 1994.

Chapter 5

Bailey, J.E. and Ollis, D.F., Biochemical Engineering Fundamentals, 1st ed., McGraw-Hill Book Company, 1977.
Biotechnology by Open Learning, "In vitro Cultivation of Microorganisms", "Bioreactor Design and Product Yield", "Energy Sources for Cells", "Product Recovery in Bioprocess Technology", Open universiteit and Thames Polytechnic, 1992.
Blanch, H.W. and Clark, D.S., Biochemical Engineering, 1st ed., Marcel Dekker, Inc.,1996.
Claudia F-W., Pfizer's penicillin pioneers, www.tcetoday.com, 2010.
Crueger, W. and Crueger, A., Biotechnology - A Textbook of Industrial Microbiology, 2nd ed., 1990.
Doran, P. M., Bioprocess Engineering Principles, 2nd ed., Elsevier, 2012.
Glazer, A. N. and Ikaido, H., Microbial Biotechnology: Fundamentals of Applied Microbiology, 2nd ed., Cambridge University Press, 2007.
Higgins, J., Best, D.J. and Jones, J., Biotechnology, 1st ed., Blackwell Scientific Publications, 1985.
Ko, Y. W., Biopharmaceuticals R & D Process, The Korean Society for Biotechnology and Bioengineering(KSBB) ed. 4, World Science, 2015.
Kyle, R.A. and Shampo, M.A.,(1997), Theodor Svedberg and the Ultracentrifuge, Mayo Clinic Proc., 72, 9, 830.
Lee, J. M., Biochemical Engineering, Prentice-Hall Inc., 1992.
McNeil, B. and Harvey, L. M., Fermentation, 1st ed., IRL Press, 1990.
Pirt, S.J., Principles of Microbes and Cell Cultivation, 1st ed., John Wiley & Sons, 1975.
Prave P., Faust, U., Sittig, W. and Sukatsch, D.A., Basic Biotechnology: VCH, 1987.
Ratledge, C. and Kristiansen, B., Basic Biotechnology, Cambridge University Press, 3rd ed., 2006.
Rao, D.G., Introduction to Biochemical Engineering, McGraw-Hill Co., 3rd ed., 2007.
Sahm, H., Antrankian, G., Stahmann, K-P. and Takors. R, Industrielle Mikrobiologie, Springer, 2013.

Schuler, M. L. and Kargi, F., Bioprocess Engineering, 1st ed., Prentice-Hall, Inc., 1992.

Scragg, A., Biotechnology for Engineers, 1st ed., Ellis Horwood Limited, 1988.

Smith J. E., Biotechnology, 4th ed., Cambridge University Press, 2005.

Song, J. Y., Biopharmaceuticals, Hongreung Science Pub., 2011.

Stanbury, P. F. and Whitaker, A., Principles of Fermentation Technology, 1st ed., Pergamon press, 1984.

Thieman W. J. and Palladino, M.A., Introduction to Biotechnology, Pearson, 2004.

Verachtert H. and Mot R. D., Yeast; Biotechnology and Biocatalysis, Dekker, 1984.

Walsh, G., Pharmaceutical Biotechnology – Concepts and Applications, John Wiley & Sons Ltd., 2007.

Wang, D. I. C., Cooney, C. L., Demain, A. L., Dunnill, P., Humphrey, A. E. & Lilly, M. D., Fermentation and Enzyme Technology, 1st ed., John Wiley & Sons, 1979.

Wiseman, A., Principles of Biotechnology, 2nd ed., Surrey University Press, 1988.

Chapter 6

Bailey, J.E. and Ollis, D.F., Biochemical Engineering Fundamentals, 1st ed., McGraw-Hill Book Company, 1977.

Biotechnology by Open Learning, Bioreactor Design and Product Yield, Open universiteit and Thames Polytechnic, 1992.

Blanch, H. W. and Clark, D. S., Biochemical Engineering, 1st ed., Marcel Dekker, Inc., 1996.

Crueger, W. and Crueger, A., Biotechnology – A Textbook of Industrial Microbiology, 2nd ed., 1990.

Doran, P. M., Bioprocess Engineering Principles, 2nd ed., Elsevier, 2012.

Glazer, A. N. and ikaido, H., Microbial Biotechnology: Fundamentals of Applied Microbiology, 2nd ed., Cambridge University Press, 2007.

Higgins, J., Best, D.J. and Jones, J., Biotechnology, 1st ed., Blackwell Scientific Publications, 1985.

Lee, J. M., Biochemical Engineering, Prentice-Hall Inc., 1992.

McNeil, B. and Harvey, L. M., Fermentation, 1st ed., IRL Press, 1990.

Pirt, S. J., Principles of Microbes and Cell Cultivation, 1st ed., John Wiley and Sons, 1975.

Schuler, M. L. and Kargi, F., Bioprocess Engineering, 1st ed., Prentice-Hall, Inc., 1992.

Scragg, A, Biotechnology for Engineers, 1st ed., Ellis Horwood Limited, 1988.
Stanbury, P.F. and Whitaker, A., Principles of Fermentation Technology, 1st ed., Pergamon press, 1984.
Wang, D. I. C., Cooney, C. L., Demain, A. L., Dunnill, P., Humphrey, A. E. and Lilly, M. D., Fermentation and Enzyme Technology, 1st ed., John Wiley & Sons, 1979.
Wiseman, A., Principles of Biotechnology, 2nd ed., Surrey University Press, 1988.

Chapter 7

Buchholz, K., Biocatalysts and Enzyme Technology, Wiley-VCH, 2005.
Chibata I., Tosa, T., and Sato, T.(1986), Immobilized cells and enzymes, Journal of Molecular Catalysis, 37, 1-24.
Fessner, WD. Biocatalysis from Discovery to Application, Springer-Verlag, 2000.
Guisan, J.M., Immobilization of Enzymes and Cells, 3rd ed., Humana Press, 2013.
Lee, J. H., Kim, S. B., Kang, S. W., Song, Y. S., Park, C., Han, S. O., and Kim, S. W.(2011), Biodiesel production by a mixture of *Candida rugosa* and *Rhizopus oryzae* lipases using a supercritical canbon dioxide process, Bioresource Technology, 102,2, 2105-2108.
Lee, J. H., Kim, S. B., Park, C., and Kim, S. W.(2010), Effect of buffer mixture system on the activity of lipases during immobilization process, Bioresource Technology, 101. 1, s66-s70.

Lee, J. M., Biochemical Engineering, Prentice-Hall Inc., 1992.
Minteer, S. D., Enzyme Stabilization and Immobilization: methods and protocols, Humana Press, 2011.
Palmer, T. Understanding Enzymes, 4th ed., Prentice Hall/Ellis Horwood Ltd., 1995.
Price, N. C. and Stevens, L., Fundamentals of Enzymology – The cell and molecular biology of catalytic proteins, 3rd ed., Oxford University Press, 2000.
Pundir, C. S., Enzyme Nanoparticles, William Andrew Publishing, 2015.
Purich, D., Enzyme Kinetics: Catalysis & Control, Elsevier, 2010.
Shuler M. L. and Kargi, F., Bioprocess Engineering – Basic Concepts, 2nd ed., Prentice-Hall Inc., 2002.
Walsh, G and Headon, D. Protein Biotechnology, 1st ed., John Wiley & Sons, 1994.
Willaert, R. G., Baron, G. V., and Backer, L. D., Immobilised Living Cell Systems – Modelling and Experimental Methods, John Wiley & Sons, 1996.

Chapter 8

Asenjo, J. A., Separation Processes in Biotechnology, Marcel Dekker Inc., 1990.

Belter, P. A., Cussler, E.L., and Hu, W-S., Bioseparations – Downstream Processing for Biotechnology, John Wiley & Sons, 1988.

Forciniti, D., Industrial Bioseparations: Principles and Practice, Blackwell Publishing Ltd., 2008.

Ghosh, R., Principles of Bioseparations Engineering, World Scientific Publishing Co., 2006.

Harrison, R. G., Todd, P., Rudge, S. R., and Petrides, D. P., Bioseparations Science and Engineering, Oxford University Press, 2003.

Harrison, R. G., Bioseparation Basics, AIChE(American Institute of Chemical Engineers), www.aiche.org/cep, October 2014.

Kim W.-S.(2007), Principles and applications of crystallization technology, Korean Industrial Chemistry(KIC) News,10, 5, 9–24.

Ladisch, M.R., Bioseparations Engineering – Principles, Practice, and Economics, John Wiley & Sons Inc., 2001.

Lee H. Y., Yoon, W. B., Song, S. H., Ahn, J. H., Park, S. J., Lee, K. Y., Separation and Purification Technology for Food and Bioindustry, The Korean Society for Biotechnology and Bioengineering(KSBB), ed. 2, World Science, 2012.

Lydersen, B. K., D'Elia N. A. and Nelson, K. L., Bioprocess Engineering: Systems, Equipment and Facilities, John Wiley & Sons Inc., 1994.

Schugerl, K., Solvent Extraction in Biotechnology – Recovery of Primary and Secondary Metabolites, Springer-Verlag, 1994.

Scopes, R.K., Protein Purification – Principles and Practice, Springer-Verlag, 3rd ed., 1994.

Zhang, S., Cao, X., Chu, J., Qian, J., Zhuang, Bioreactors and bioseparation, Advances in Biochemical Engineering/Biotechnology, 122, 105-150, Scheper T. ed., Springer, 2010.

Chapter 9

Bangkok Post, "Drug resistance grows menacingly", 2015-12-21.

BioINdustry, 74, Therapeutic vaccine, Biotechnology Policy Research Center(BPRC), 2013.

BioINdustry, 91, Status and prospects on the global market of In Vitro Diagnostics, Biotechnology Policy Research Center(BPRC), 2015.

BioINdustry, 220, Reports on bioindustry trends, Biotechnology Policy Research

Center(BPRC), 2015.

Bogner, E. and Holzenburg, A., New Concepts of antiviral theraphy, Springer, 2006.

Cheng, C.-M., Kuan, C.-M., and Chen, C.-F., In-Vitro Diagnostic Devices – Introduction to Current Point-of-Care Diagnostic Devices, Springer International Pub. Switzerland, 2016.

Cho, J. J., Research trends and prospects of gene therapeutics, BioIN Expert Report, 9, Biotechnology Policy Research Center(BPRC), 2015.

Ganapathy S., Biopharmaceutical Production Technology, John Wiley & Sons, Inc., 2012.

Gee, A., Cell Therapy – cGMP Facilities and Manufacturing, Springer-Verlag U.S., 2009.

Giacca M., Gene Therapy, Springer-Verlag Italy, 2010.

Go, H. S., How do you see the biosimilars in the wind?, Korea Investers Service(KIS) Credit Monitor, Special Report, 13-35, 2012.

GyeongGi Bio Insight, 4, 1, In Vitro Diagnostics, Gyeonggi Institute of Science & Technology Promotion(GSTEP), 2015.

Ho, R. J. Y. and Gibaldi, M., Biotechnology and Biopharmaceuticals, John Wiley & Sons, Inc., 2003.

Jeong, H. M., Research trends on new drug development based on stem cell, BioINpro, 16, 1-13, Biotechnology Policy Research Center(BPRC), 2015.

Jin, T. E., Bioresources, Core materials of bioeconomy period, Science & Technology Policy, 25, 11, 20-25, Science and Technology Policy Institute(STEPI), 2015.

Kang, K. S., Research trend for application of stem cell, BioINpro, 16, 1-11. Biotech. Policy Research Center(BPRC), 2015.

Kang, S. W. and Kim, S. W.(2001), Current status and prospect on biotechnology, Korean Industrial Chemistry News, 4(1), 34-44.

Kim, B. C., Personal customized medicine through molecular diagnosis – Development of targeted anticancer agents, BioIn Special WebZine, 33, Biotechnology Policy Research Center(BPRC), 2013.

Kim, M.-J., Lee, S.-W., Lee D.-K., Park, J.-E., Kang, J.-Y., Park, I.-H., Shin, H.-S., and Ha, N.-J.(2013), Antibiotic resistant patterns and DNA fingerprint analysis of *Acinetobacter baumannii* from clinical isolates, Yakhak Hoeji, 57, 2, 132-138.

Kim, T-U., Biopharmaceuticals, Status and prospect of biosimilar development, BioIn Expert Report, 9, Biotechnology Policy Research Center(BPRC), 2015.

Kim, Y. S. and Yoo, S. S., Research trend of gene therapeutics, BioIn Expert Report, 5, 1-14, Biotechnology Policy Research Center(BPRC), 2014.
Ko, Y. W., Biopharmaceuticals R & D Process, The Korean Society for Biotechnology and Bioengineering(KSBB) ed. 4, World Science, 2015.
koreaBio, 2010 koreaBio Report III, Cell therapeutics: Beginning of new technology, 1-13, Korea Biotechnology Industry Organization, 2010
Korea Centers for Disease Control and Prevention, Report on establishment of collection system about antibiotic resistant data from hospitals, 2010.
Lancini, G. and Parenti, F., Antibiotics – An Integrated View, Springer-Verlag, 1982.
Largent, M. A., Vaccine, Johns Hopkins University Press, 2012
Lee, N. G.(2015), Immunostimulator as vaccine-based technology, Molecular and Cellular Biology Newsletter, 5, 1-5.
Ministry of Science, ICT and Future Planning(MSIP), Biotechnology 2013 & 2015.
National Cancer Information Center, http://www.canaer.go.kr
Oh, I. H., Trends and prospects of stem cell therapeutics, BioIn Expert Report, 9, Biotechnology Policy Research Center(BPRC), 2015.
Prugnaud, J.-L. and Trouvin, J.-H. (eds.), Biosimilars – A New Generation of Biologics, Springer-Verlag France, 2013.
Saltzman, W. M., Biomedical Engineering – Bridging Medicine and Technology, Cambridge University Press, 2015
Song, J. Y., Biopharmaceuticals, Hongreung Science Pub, 2011.
Spencer, P. and Holt W., Anticancer Drugs: Design, Delivery and Pharmacology, Nova Science Pub., Inc., 2009.
Sung, B. R.(2012), Recent trends of vaccine research, Bioin Special, WebZine, 30, 1-22. Biotech. Policy Research Center(BPRC).
Trade Brief, 59, Activation of the Nagoya Protocol and industry impact, Institute for International Trade, 2014.
Trends on technology development of biopharmaceuticals (Gene therapeutics), Report on BT Technology Trend, 120, Biotechnology Policy Research Center(BPRC), 2010.
Trends on technology development of biopharmaceuticals (Antiviral agents), Report on BT Technology Trend, 121, Biotechnology Policy Research Center(BPRC), 2010.
Walsh, G., Biopharmaceuticals – Biochemistry and Biotechnology, 2nd ed., John Wiley & Sons Ltd., 2003.

Chapter 10

Baskar, C., Baskar, S., and Dhillon, R. S. (eds.), Biomass Conversion, Springer-Verlag Germany, 2012.

Brenes, M. D.(ed.), Biomass and Bioenergy: New Research, Nova Science Pub., 2006.

Cha, W. O.(2007), Trends on biodiesel industry, News & Information for Chemical Engineers(NICE), 25, 6, 609-613.

Chang H. N. Kim, N-J., Kang, J., and Jeong, C. M.(2010), Biomass-derived volatile fatty acid platform for fuels and chemicals, Biotechnol. Bioprocess Eng.,15,1,1-10.

Dumeignil, F., Dibenedetto, A., and Aresta, M. (eds.), Biorefinery: From Biomass to Chemicals and Fuels, De Gruyter, 2012.

Fanchi, J. R., and Fanchi, C. J., Energy in the 21st Century, 3rd ed., World Scientific Pub., 2013.

Hwang, I. T., Hwang, J. S., Lim, H. K. and Park, N.-J.(2010), Biorefinery based on weeds and agricultural residues, Korean Journal of Weed Science, 30(4), 340-360.

Kim, E. S., Hyun, S. E., Lee, K. S., and Kim, K. S.(2014), Research trend on ionic liquid, NICE, 32, 1, 56-64.

Kim, J.-H., Kim, Y.-H., Kim, S.-K., Kim, B.-W., Nam. S.-W.(2011), Properties and industrial applications of seaweed polysaccharides-degrading enzymes from the marine microorganisms, Korean J. MIcrobiol. Biotechnol., 39, 3, 189-199.

Kim, S.B., Park, C., and Kim S.W.,(2014), Process design and evaluation of production of bioethanol and beta-lactam antibiotic from lignocellulosic biomass, Bioresource Technology, 172, 194-200.

Kim, S. K., Hwang, H. J., Kim, J. D., Ko, E. H., Choi, J. S., and Kim, J.-S.(2012), Usefulness of freshwater alga water-net (*Hydrodictyon reticulatum*) as resources for production of fermentable sugars, Korean Journal of Weed Science, 32(2), 85-97.

Lee, J. H., Kim, S. B., Kang, S. W., Song, Y. S., Park, C., Han, S. O., and Kim, S. W.(2011), Biodiesel production by a mixture of *Candida rugosa* and *Rhizopus oryzae* lipases using a supercritical canbon dioxide process, Bioresource Technology, 102, 2, 2105-2108.

Lee J. H. and Kim, S. W.(2007), Biodiesel production by enzymatic method, NICE, 25, 6, 617-619.

Lee, J. S.(2007), Biodiesel production by chemical catalyst, NICE, 25, 6, 613-617.

Lee, J. S. and Kim, S. W.(2009), Conversion of crude glycerol to the value-added products, NICE, 27, 4, 434-440.

Lee, Y. W.(2007), Production technology of biodiesel fuel by supercritical fluid, NICE, 25, 6, 620-625.

Liguori, R. and Faraco, V.(2016), Biological processes for advancing lignocellulosic waste biorefinery by advocating circular economy, Bioresource Technology, 215, 13-20.

Ministry of Science, ICT and Future Planning(MSIP), Biotechnology, 2013 & 2015.

NREL(National Renewable Energy Laboratory), http://www.nrel.gov

Oh, Y.-K. and Na, J.-G.(2015), Microalgal biodiesel production process, Korean Industrial Chemistry(KIC) News, 18, 3, 1-14.

Shin, S. C.(2007), Quality control of biodiesel and its activation plan, NICE, 25, 6, 626-631.

Special Themes, TPA Production Technology from Biomass, Trend in White Biotech, 71, 30-43, 2015.

Thapa, L. P., Lee, S. J., Yang, X., Lee, J. H., Choi, H. S., Park, C., and Kim, S. W. (2015), Improved bioethanol production from metabolic engineering of *Enterobacter aerogenes* ATCC 29007, Process Biochemistry, 50,12, 2051-2060.

The Korean Academy of Science and Technology(KAST), Status and Challenges of Bio-based Fuel and Chemical Industry, Research Report 50, 2008.

Yang, X., Choi, H. S., Park, C., and Kim, S. W.(2015), Current states and prospects of organic waste utilization for biorefineries, Renewable and Sustainable Energy Reviews, 49, 335-349.

Yang X., Lee, S. J., Yoo, H. Y., Choi, H. S., Park, C., and Kim, S. W.(2014), Biorefinery of instant noodle waste to biofuels, Bioresource Technology, 159, 17-23.

Yoon, Y. Y.(2014), Domestic biomass utilization and promotion, World Agriculture,162, Global agro-food industry trend, 1-25.

Wall, J. D., Harwood, C. S., and Demain, A., Bioenergy, ASM Press, 2008.

INDEX